AF531309

COMPUTATIONAL PHYSICS

COMPUTATIONAL PHYSICS

By

D.K. Jha

DISCOVERY PUBLISHING HOUSE PVT. LTD.
NEW DELHI-110 002

Published by:
Namit Wasan

DISCOVERY PUBLISHING HOUSE PVT. LTD.
4383/4B, Ansari Road, Darya Ganj
New Delhi-110 002 (India)
Phone : +91-11-23279245; 23253475; 43596065
E-mail : discoverybooksindia@gmail.com
discoverypublishinghouse@gmail.com
namitwasan9@gmail.com
web : www.discoverypublishinggroup.com

Edition: **2020**

ISBN: 978-81-8356-451-9

Computational Physics

Printed at:
Infinity Imaging Systems
Delhi

Preface

The present title *"Computational Physics"* has been written with the aim to provide readers the detailed information experimentally accessible computational approaches based on detailed atomic models. This book is organized so that non specialists as well as more advanced users can benefit. It can serve as both as introductory text to computational sciences making it useful for undergraduate and postgraduate students, and also a reference source for active researchers in the field. Efforts have also been made to compile a comprehensive but reasonably concise review of relevant theoretical and computational methods that is self contained. In addition, the book provides the needed information for calculating the interpreting experimentally observed properties on the basis of results generated by centre simultations.

There can be no claim to originality except in the manner of treatment and much of the information has been obtained from the books and scientific journals available in the different libraries.

The author expresses his thanks to his friends and colleagues whose continue inspirations have initiated him to bring out this book.

The author is painfully aware of the shortcomings, errors and misprints that have crept in, and shall be grateful to receive suggestion for improvement of the next edition from all the readers.

The author expresses his gratitude to Mr. Wasan and staff of M/s Discovery Publishing House Pvt. Ltd. for their whole hearted co-operation in the publication of this book.

—Author

CONTENTS

1

INTRODUCTION

Some bacteria, such as *Escherichia coli,* swim by spinning long helical flagella. Each cell has multiple flagella, all of which have the same handedness. When the flagella are spun in the counterclockwise (CCW) direction, they come together to from a bundle that propels the cell through the fluid. The motor that is responsible for flagella rotation is reversible. When spun in the clockwise (CW) direction, the flagella fly apart and the cell undergoes a tumbling motion. Addition of a chemical attractant causes the cell to suppress tumbling when moving toward this food source. One of the proteins in the signaling pathway is CheY. The binding of phosphorylated CheY to the portion of the motor located within the cytoplasm promotes CW rotation. To model motor reversals, the protein complex that forms the rotor is assumed to exist in two distinct conformational states that correspond to CW and CCW rotation. The binding of CheY decreases the free energy of the CW state, while at the same time increasing the free energy of the CCW by an equivalent amount. To capture this effect, the free energy of the rotor is assumed to have the following form

$$G(x)=4\Delta G_{nb}\left(\frac{x^4}{4}-\frac{x^2}{2}\right)-\frac{1}{2}\Delta Gx, \qquad ...(1)$$

where x is an appropriate reaction coordinate, AG = G(–1) – G(l) is the free energy difference between the CW and CCW states, and ΔG_{nb} = G(0) – G(–1) is the free energy difference between the transition state and either CW or CCW state when the chey concentration is such that *(i.e.)* there is no bias toward CCW or CW rotation.

To model the chemical kinetics of chey, we assume that binding of chey to the motor is a two-rate processes with a single bindking site being either empty or occupied. Let PE (t) be the probability that the site is empty at time it and PO (t) the probability that the site is occupied. The probabilityes satisfy the following set of coupled equations:

$$\frac{dp_E(t)}{dt} = - k_{on} p_E(t) + k_{off} p_O(t), \tag{2}$$

$$\frac{dp_O(t)}{dt} = - k_{off} p_O(t) + k_{on} p_E(t), \tag{3}$$

where k_{off} is the dissociation rate constant and k_{on} is the rate at which CheY binds to the motor. From the law of mass action, k_{on} should be proportional to the CheY concentration. That is, k_{on} = k'_{on}[CheY], where k'_{on} is a second-order rate constant and the brackets stand for concentration. If the concentration of CheY is held constant, then p_E and p_O will relax to their equilibrium values. These are found by solving (2) and (3) with the time derivatives set equal to zero. Doing this yields p_O = [CheY]/(K_d + [CheY]) where $K_d = k_{off}/k'_{on}$ = 9.1 μM is the dissociation constant.

There are approximately 26 binding sites on the motor, and the average number of occupied sites is $26p_O$. In the absence of CheY, $\Delta G = \Delta G_O = 14k_BT$ and at saturating concentrations of CheY, $\Delta G = \Delta G_\infty = -9k_BT$. Therefore, the change in ΔG from low to high CheY concentrations is $\Delta\Delta G = \Delta G_0 - \Delta G_\infty = 23k_BT$, where the symbol AAG indicates that we are talking about a change in the value of AG. Thus each CheY contributes roughly $0.88k_BT$ toward changing the relative free energy of the CW and CCW states. These considerations lead to the following expression for ΔG:

$$\Delta G = \Delta G_0 - \Delta\Delta G \frac{[\text{CheY}]}{K_d + [\text{CheY}]}. \tag{4}$$

Graphs of the free energy at various CheY concentrations with two minima located roughly at $x = \pm 1$. Generally, we shall measure distance x in nanometers (nm) and force in piconewtons (pN). In these units k_BT = 4.1 pN-nm at room temperature (T = 298 K). The force that arises from changes in free energy is $-\partial G/\partial x$. Therefore, the force vanishes at the minima. Additionally, if the conformation of the rotor is slightly displaced from either minimum, it experiences a force that moves it back toward that minimum. The force also vanishes at the local maximum located near $x = 0$. However, when the conformation of the rotor is slightly displaced from the origin, the force acts to move away from the rotor at $x = 0$ and toward one of the two minima. Thus, we expect the rotor to spend most of its time near the minima. To surmount the energy barrier between the minima requires a substantial thermal fluctuation.

The reaction coordinate $x(t)$, which determines the state of the rotor, can take on values anywhere between $\pm\infty$. Clearly, we cannot use an infinite interval in our numerical algorithm. However, since $G(x) \to \infty$ as $x \to \pm\infty$, there is a strong restoring force that drives the reaction coordinate back toward the origin when $|x|$ is large. This means that the probability of finding $x(t)$ at distances far from the origin is small, and ignoring large values of $|x|$ will not significantly affect our numerical solutions. For the parameters we shall consider, the interval $(-2,2)$ is wide enough to ensure numerical accuracy. In practice, an appropriate interval can be determined by successively enlarging the length until the numerical results no longer change appreciably. At $x = \pm 2$ we enforce reflecting boundary conditions. The diffusion equation for this process is

$$\frac{\partial p}{\partial t} = D\frac{\partial}{\partial x}\left(\frac{\partial G(x)}{\partial x}\frac{p}{k_B T} + \frac{\partial p}{\partial x}\right) \tag{5}$$

To find the equilibrium distribution for $p(x, t)$, the above equation is solved with $\partial p/\partial t = 0$. This yields

$$p_{eq}(x) = \frac{\exp\left(-\dfrac{G(x)}{k_B T}\right)}{\int_{-\infty}^{\infty}\exp\left(-\dfrac{G(y)}{k_B T}\right)dy}. \tag{6}$$

The diffusion coefficient D in (5) represents an effective diffusion coefficient for the reaction coordinate that includes many microscopic effects. For all the results presented below, $D = 70\ \mathrm{nm}^2/\mathrm{s}$ and $\Delta G_{nb} = 5k_B T$. These values were chosen to be consistent with experimental observation that at 14μM of CheY, motor reversals occur at an average rate of 2/s. The bistable nature of the system is clearly evident. The time series can be used to produce histograms of the reaction coordinate. An approximation for the distribution $p_n^{(s)}$ is constructed by dividing the number of points in each bin of the histogram by the total number of points in the time series. Then we estimate $p(x_n) \approx p_n^{(s)}/\Delta x$.

In the discussion of Markov chains and diffusion in the previous chapter, we encountered the idea of a waiting or first passage time. This is a very important mathematical concept that comes up in many different biological contexts. At a CheY concentration of 14 μM the motor reverses roughly twice per second, and there is no bias toward CW or CCW rotation. To compute the switching rate, we must compute the average time for the system located at the reaction coordinate x to surmount the energy barrier at $x = 0$. To this end, the reaction coordinate is started *at* $x = -1$ at $t = 0$ with an absorbing boundary at $x = 0$. To generate this figure, 61 grid points were used. Note that the probability of finding the particle in the interval $(-\infty, 0)$ is continuously decreasing, due to the absorbing boundary. This probability can be used to determine the first passage time density $f(t)$ through the relation

$$f(t) = -\frac{d}{dt}\text{Prob}=[-\infty < x(t) < 0] = -\frac{d}{dt}\int_{-\infty}^{0} p(x,t)dx$$
$$\approx -\frac{d}{dt}\sum_{n=1}^{M} p_n(t) = p_M F_{M+1/2}^{abs}, \quad \text{...(7)}$$

where the last equality and the absorbing boundary condition. Therefore, the numerical algorithm is well suited for computing first passage time densities. The solid line is the numerical result. The dashed line is the Kramers approximation, which assumes that the process has an exponential distribution with mean first passage time

$$MFPT \approx \frac{k_B T \pi}{D\sqrt{G''(-1)\,|\,G''(0)\,|}}\exp\left(\frac{G(0)-G(-1)}{k_B T}\right)$$
$$= \frac{k_B T \pi}{D\sqrt{32\,\Delta G_{nb}}}\exp\left(\frac{\Delta G_{nb}}{k_B T}\right). \quad \text{...(8)}$$

Note that the most significant factor in determining the mean first passage time is ΔG_{nb}. The validity of the Kramers approximation depends on $\Delta G_{nb} >> k_B T$. The first passage time distribution is not exponentially distributed, since it must be equal to zero at $t - 0$. However, if we ignore this very short initial time interval, the distribution is approximated reasonably well with an exponential. Using the numerical distribution to compute the mean first passage time, we obtain MFPT = 0.253 s, and using Kramers approximation (8) we find MFPT = 0.236s. An exact

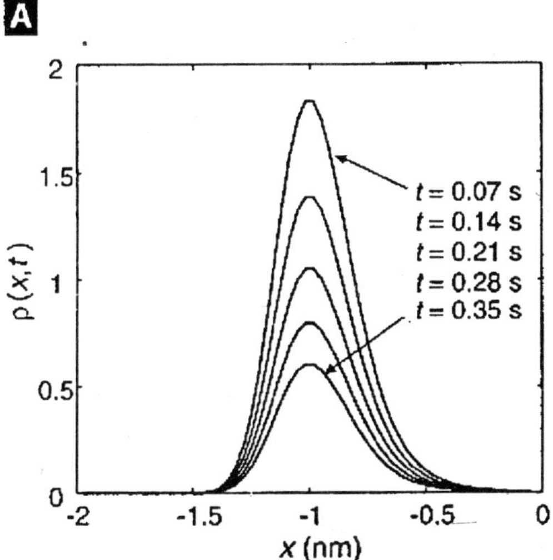

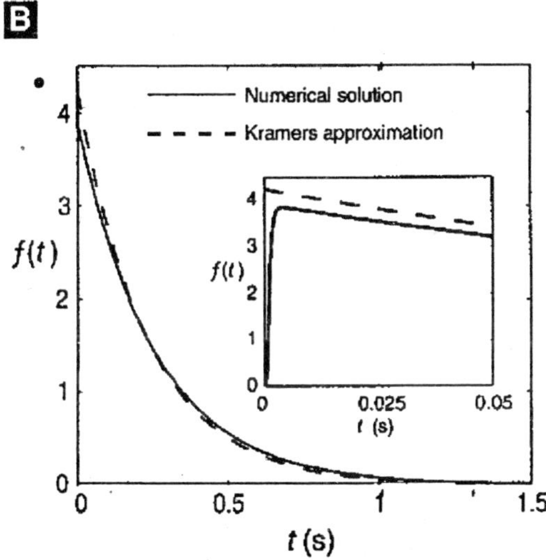

Fig. 1.1. Numerical results of the first passage time problem. (A) Time evolution of the probability density. (B) Probability density of the first passage time.

expression gives MFPT = 0.259 s. The agreement between this value and the numerical result given above provides evidence that the algorithm is faithfully reproducing the dynamics of the system. The switching rate is l/(2 MFPT) = 1.98/s, where the factor of 1/2 comes from the fact that half the time the reaction coordinate surmounts the barrier, it falls back into the well from which it started. This justifies our choice of D and $AG_n b$.

To test this simple model, we compute the CW bias as a function of CheY concentration. The CW bias is the fraction of time that the motor spends rotating in the CW direction. This can be computed by integrating (6) over the interval (-co, 0). That is,

$$\text{CW bias} = \frac{\int_{-\infty}^{0} \exp\left(-\frac{G(x)}{k_B T}\right) dx}{\int_{-\infty}^{\infty} \exp\left(-\frac{G(y)}{k_B T}\right) dy} \qquad \text{...(9)}$$

Figure 1.2 shows a comparison of the model's predictions to the experimental data of Scharf et al. The agreement between the data and the theoretical curve provides evidence to support the model's validity.

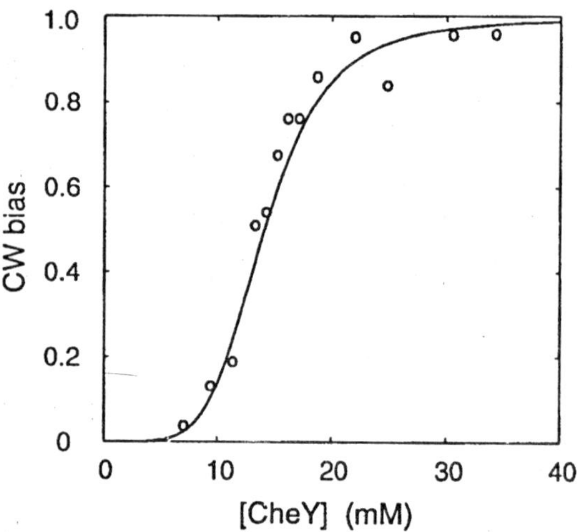

Fig. 1.2. The CW bias as function of CheY concentration.

Motor Driven by a "Flashing Potential"

The following process, called the *flashing ratchet*, is a paradigm for molecular motors. Imagine a protein driven by alternating its exposure to two potential energy profiles: V_1 (solid line) and V_2 (dashed line), as shown in Figure 1.3. The first potential is a piece wise linear asymmetric sawtooth potential, while the second potential is a constant. Thus, in the first potential, the protein is localized near a local minimum, while in the second potential the

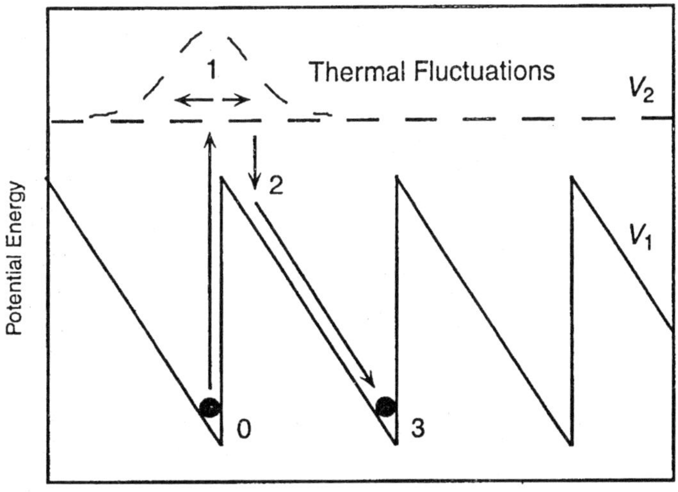

Fig. 1.3. The flashing ratchet. In the first potential, the protein is localized near a local minimum. Alternatively, the protein is free to diffuse.

protein diffuses freely. While in either potential, the motion of the particle is given simply by $\zeta dx/dt = -dV/dx + fB(t), i = 1,2$. Switching between the potentials is governed by a chemical reaction (vertical arrows), which occurs with rate k.

Clearly, if the sawtooth potential is symmetric, the average displacement of the protein must be zero. However, in the case of the asymmetric potential shown in Figure 1.3, the protein moves on the average to the right, although all steps are reversible. This phenomenon can be easily understood when the following inequalities are valid:

$$\frac{k_B TL}{FD} \ll \frac{1}{k} \ll \frac{L^2}{D} \qquad \text{...(10)}$$

Here L is the wavelength of the sawtooth potential, and $F = -dV_1/dx$ is the slope of the first potential, i.e., the force driving the protein to the right in the first potential. The corresponding drift rate of the protein is $F/\zeta = FD/k_B T$. The order of magnitude of the time for the protein to drift into a local minimum is $k_B TL/FD$. The diffusion in the first potential and the protein's deviations from a local minimum can be neglected if the slope of the sawtooth potential is very steep. Quantitatively, this means that $k_B T \ll FL$ (or $k_B TL \ll FL^2$; compare with (10).

The first inequality (10) means that the protein reaches a local minimum of the sawtooth potential well before this potential switches off. The second inequality (10) indicates that when the protein diffuses freely, it rarely can move farther than distance L before the sawtooth potential is switched back on: the mean time between "flashes" of the potential, l/k, is much less than the characteristic diffusion time L^2/D. Thus by the time the sawtooth potential is switched off, the protein is in a local minimum. When the sawtooth potential is switched off, the protein diffuses with equal probability to the left and to the right, and will not diffuse very far compared to the period of the potential. If the protein diffuses to the left, then by the time the sawtooth potential is switched back on, the protein is in the *basin of attraction* of the same local minimum it started from. When the potential is on, the protein returns to the starting point of the cycle. However, if the diffusion to the right took place, the protein is in the basin of attraction of the next local minimum to the right. Thus, the protein either does not move, or moves the distance L to the right, with equal probability. Said another way, the diffusion in the flat potential can be viewed as a spreading Gaussian distribution. The asymmetry of the sawtooth potential, when it is switched on, cuts a larger portion of the distribution into the next domain of attraction. The corresponding average rate of motion is $(V) - Lk/4$ (the mean duration of a cycle is $21k$, and on average, the system steps a distance L per two cycles).

We considered the highly peculiar sawtooth potential with one of the slopes being infinitely steep. For more regular smooth potentials, the velocity of the flashing ratchet is computed analytically in the so-called fast and slow flashing limits. In both of these limits, the protein advances very slowly. If the flashing is too fast, the protein does not have time to reach the local minimum, and the effect of asymmetry is lost. The protein is effectively exposed to the average potential, which does not support any steady movement. On the other hand, if the flashing is too slow, the freely diffusing protein moves too far from a local minimum, the information about its initial position is lost, and the average displacement becomes very small. The mean velocity of the flashing ratchet reaches a maximum $\approx Lk/4$ (for the smooth asymmetric potential, such that $\delta V \approx k_B T$) when $k \approx D/L^2$ (flashing frequency is of the same order of magnitude as the inverse time to diffuse over the potential's period).

In the general case, the average velocity of the flashing ratchet can be computed only numerically. Following the methods of the previous chapter, we can describe the ratchet by two coupled Smoluchowsky equations:

$$\frac{\partial p_i}{\partial t} = D\frac{\partial}{\partial x}\left[\frac{\partial p_i}{\partial x} + \frac{\partial V_i/\partial x}{k_B T} p_i\right] + k(-p_i + p_j),\ i=1,2,j \neq i. \quad ...(11)$$

These equations must be solved numerically on the finite domain [0, *L*] with periodic boundary conditions and normalization condition

$$\int_0^L (p_1(x,t=0) + p_2(x,t=0))dx = 1.$$

When the probability distributions achieve steady state, the net current is

$$J = -D\left[\frac{\partial^2 (p_1 + p_2)}{\partial x^2} + \frac{\partial V_1/\partial x}{k_B T} p_1\right],$$

from which the average velocity can be found: $(V) = LJ$.

It is important to realize that in the process energy is consumed from the chemical reaction that drives the switching between the two potentials. The motion down the slope of the sawtooth potential $2 \rightarrow 3$ generates heat by frictional dissipation. Finally, if a small load force directed to the left is applied to the protein, the movement slows down. The load force is equivalent to tilting the potential to the left. Thus, the flashing ratchet is able to generate force, and has all characteristics of a molecular motor. However, there is no direct correspondence of the flashing ratchet mechanism to a real motor. In what follows, we consider two simple models of actual molecular motors.

Polymerization Ratchet

Perhaps the simplest way to convert chemical energy into a mechanical force is by polymerizing a filament against a load force (see the previous chapter). Here the energy source is the free energy of binding of a monomer onto the tip of the polymer, ΔG_b. If a polymer assembles against no resistance, the polymerization velocity (elongation rate) is simply

$$V_p = L(k_{on}M - k_{off}) \quad (12)$$

where L (nm) is the size of the monomer, M (μM) the monomer concentration, and k_{on} (1/μM·s), k_{off} (1/s) are the polymerization and depolymerization rate constants, respectively.

If an object is placed ahead of the growing polymer, there are two mechanisms by which the polymer·can "push" the object: (i) by rectifying the Brownian motion of the object; (ii) by actively "pushing" against the object, i.e., a power stroke. First, we discuss the Brownian ratchet. We assume that the polymer is anchored at the left end and is perfectly rigid. The object has a diffusion coefficient $D = k_B T/\zeta$. For the moment, we neglect depolymerization ($k_{off} = 0$). In order for a monomer to bind to the end of the filament the object must open up a gap of size L by diffusing away from the tip, and remaining there for a time $\approx (k_{on}M)^{-1}$ to allow a polymerization event to take place. In the limiting case when polymerization.is much faster than diffusion, i.e., $k_{on}M >> D/L^2$, we can consider the polymerization to happen instantaneously once a gap of size L appears. Then the time for the load to diffuse a distance L is simply the mean first passage time $(T) = L^2/2D$. To cover N such intervals takes $N \cdot \tau$ time units, so the average velocity is simply $(V_p) = NL/(N{\cdot}(T)) = 2D/L$. This is the speed of an ideal Brownian ratchet. Note that by reducing the size of the monomer L, the speed of the ratchet increases, since the likelihood of a thermal fluctuation of size L increases exponentially as L decreases. However, this is true only as long as our approximation holds: $k_{on}M >> D/L^2$, or $L >> \sqrt{D/(k_{on}M)}$. For smaller values of L the polymerization reaction becomes the limiting factor, so that $V_p \approx L \cdot k_{on}M$ (cf. (12) with $k_{Off} = 0$).

We can picture the situation as shown in Figure 1.4B: the object diffuses on a "staircase" sequence of identical free energy functions $\phi(x)$, each with a step height of the monomer binding free energy $\Delta G = -k_B T \ln(k_{on}M/k_{off})$. If a load force F_L opposes the diffusive motion of the object, the potential becomes $\phi(x) - F_L x$. This corresponds to tilting the potential so that the object must diffuse "uphill,".

Including the depolymerization rate complicates the analysis considerably. However, a diffusion equation can be formulated that can be solved exactly when $k_{on}ML$, $k_{off}L << 2D/L$. In this regime, the approximate load-velocity relationship is given by the simple formula

$$V_p = L(k_{on}Me^{-f_L L/k_B T} - k_{off}). \qquad ...(13)$$

That is, the polymerization rate in (13) is weighted by a Boltzmann factor where the exponent $F_L L/k_B T$ is the work done by the load in moving the object one step distance L. The stall

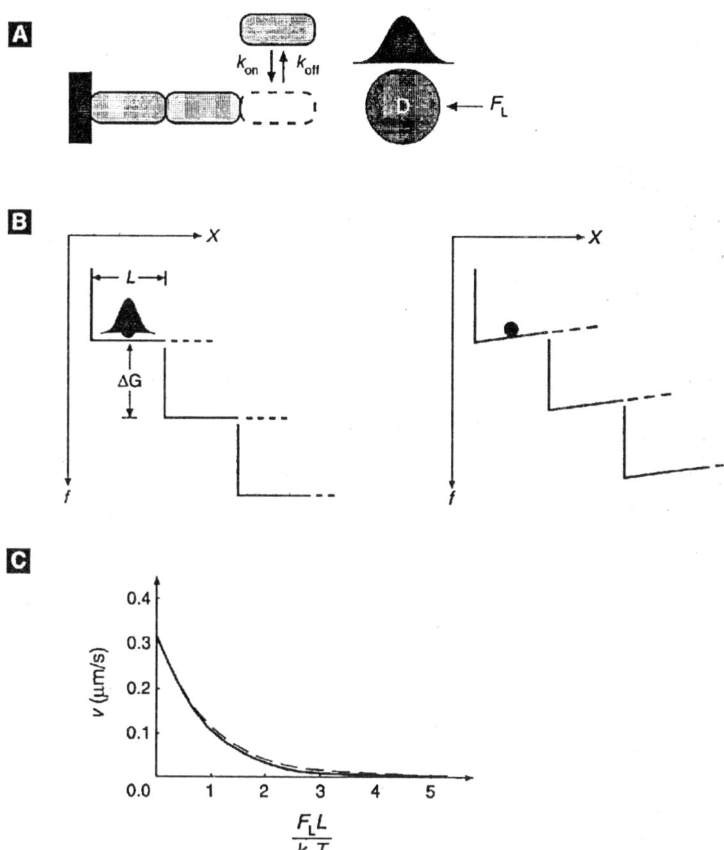

Fig. 1.4. The polymerization ratchet. (A) Monomers of length L polymerize onto the end of a growing filament with rate constants k_{on}, k_{off} (B) The energy diagram of the polymerization process. (C) The load-velocity curve for the polymerization ratchet.

load F_s is reached when the work done in moving the object a distance L *is* just equal to the free energy from the binding reaction, so that $V_p = 0$:

$$F_s = \frac{k_B T}{L} \ln\left[\frac{k_{on} M}{k_{off}}\right] \qquad ...(14)$$

Note that without depolymerization, $k_{off} \rightarrow 0$, there is no finite stall load.

Variations and elaborations on the polymerization ratchet have been used to model a variety of cellular processes, including lamellipodial protrusion, the polymerization of microtubules, the propulsion of intracellular pathogens, and the translocation of proteins.

Simplified Model of the F_0 Motor

To futher illustrate the numerical formalism, we shall examine in detail a simplified model based on the principle of the ion-driven F_0 motor of ATP synthase. This enzyme uses electrochemical energy stored in a proton motive force across the inner membrane of mitochondria to produce ATP. This will illustrate many of the principles of mechanochemical energy conversion by proteins, but is sufficiently simple to analyze analytically and numerically. It consists of two reservoirs separated by an ion-impermeable membrane. The reservoir on the left is acidic (high proton concentration) with concentration c^{acid}, and the reservoir on the right is basic (low concentration) with concentration c^{base}. The motor itself consists of two "parts": (i) a "rotor" carrying negatively charged sites spaced a distance L apart that can be protonated and deprotonated; (ii) a "stator" consisting of a hydrophobic barrier that is penetrated by an apolar strip that can allow a protonated site to pass through the membrane, but will block the passage of an unprotonated site. The height of the energy barrier blocking passage of a charge between two media with different dielectric constants $\in_1$ and $\in_1$ is $\Delta G \approx 200[(1/\in_1) - (1/\in_2)] \approx 45k_BT$. This energy penalty arises from the necessity of stripping hydrogen-bonded water molecules from the rotor sites.

Qualitatively, the motor works like this. Rotor sites on the acidic side of the membrane are frequently protonated. In this state (a nearly neutral dipole) the rotor can diffuse to the right, allowing the protonated site to pass through the membrane-stator interface to the basic reservoir. Once exposed to the low proton concentration in the basic reservoir, the proton quickly dissociates from the rotor site. In its charged state, the rotor site cannot diffuse backwards across the interface: Its diffusion is "ratcheted."

Thus a rotor site can exist in two states: unprotonated and protonated. To specify the mobility state of the rotor, we need to keep track only of the site immediately adjacent to the membrane on the acidic side. In its unprotonated state, the site adjacent to the membrane is immobilized, since it cannot pass into the stator,

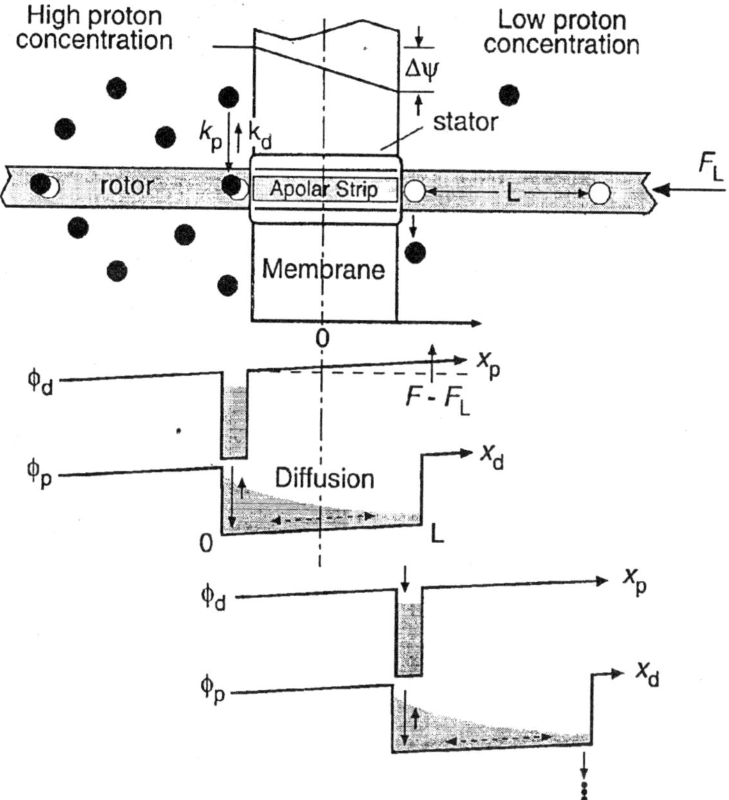

Fig. 1.5. Simplified model illustrating the principle of the F_0 motor.

nor can it diffuse to the left, since the next rotor site on the basic side of the membrane is almost always deprotonated, and cannot diffuse to the left. (Of course, this depends on the thickness of the membrane being the same as the rotor spacing; this is unrealistic, but the full model treated in the references does not have this constraint.) Thus the progress of the model can be pictured as a sequence of transitions between two potentials, as shown in Figure 1.5. When deprotonated, the rotor is immobilized in potential . and when protonated, it can move in potential ϕ_p. The effect of the load force F_L is to tilt the potentials upward, so the motion in potential ϕ_p is "uphill" (i.e., the total potential when protonated can be written as $\phi_p(x) - F_L x$).

Below, we consider two limiting cases. In the first one, diffusion is much faster than the chemical reaction rates. In the second one, the diffusion time scale is comparable to the reaction rates in the basic reservoir. The first case can be treated *analytically*, while the second one will require *numerical* simulation.

Average Velocity of the Motor in the Limit of Fast Diffusion

The model can be formulated mathematically in terms of the probability of the depro-tonated state $p_d(t)$ (non-dimensional), and the probability density of the protonated state $p_p(x, t)$ (1/nm). Here x, $0 \leq x \leq L$, is the distance of the protonated site from the interface between the acidic reservoir and the membrane. The model equations have the form

$$\frac{dp_d}{dt} = \text{net deprotonated spatial flux + net reaction flux}$$

$$= j_{x_d} + J_\xi \qquad ...(15)$$

$$\frac{\partial p_p}{\partial t} = \text{net protonated spatial flux — net reaction flux}$$

$$= j_{x_p} + J_\xi \qquad ...(16)$$

where

$$J = 0,$$

$$J_{x_p} = D\frac{\partial}{\partial x}\left(\frac{\partial p_p}{\partial x} - \frac{F_1 - F_L}{k_B T} p_p\right),\ J_{x_p}(0) = J_{xp}(L) = 0,$$

$$J_\xi = \text{deprotonation at acidic reservoir}$$

$$+ \text{ deprotonation at basic reservoir}$$

$$- \text{ net protonation at both reservoirs}$$

$$= k_d p_p(0) + k_d p_p(L) - k_p p_d.$$

Protonation rates are proportional to hyrogen ion concentration on either side of the membrane, and the net protonation at both reserviors is $k_p = k_p c^{acid} + k_p c^{base}$. Note that the rates of protonation and deprotonation have the dimensions k_p (1/s) and k_d (nm/s), respectively. We assume that the deprotonation rates are the same at both reservoirs.

First we nondimensionalize the model equations using the rescaled coordinate $(x/L) \to x$, the rescaled time $(k_d t/L) \to t$, and the following dimensionless parameters:

1. Ratio of reaction to diffusion time scales: $\Lambda = D/k_d L$.
2. Net work done in moving the rotor a distance L; $w = (F_1 - F_L)L/k_B T$.
3. Equilibrium constant: $k = k_p L/k_d$.

Here $F_I\, e\Delta\psi/L$ is the electrical driving force, which is assumed to be constant. Substituting these variables and parameters into equations (15) and (16) gives

$$\frac{dp_d}{dt} = -kp_d + p_p(0) + p_p(1), \qquad ...(17)$$

$$\frac{\partial p_p}{\partial t} = kp_d - p_p(0) - p_p(1) + \Lambda \frac{\partial}{\partial x}\left(\frac{\partial p_p}{\partial x} - wp_p\right), \qquad ...(18)$$

where x, t *are* now the nondimensional coordinate and time, respectively.

In many situations it turns out that diffusion is much faster than the chemical reaction rates: $\Lambda >> 1$. (At $D \approx 10^7$ nm²/s, $L \approx 10$ nm, $k_d \approx 10^3$ nm/s, the order of magnitude of the parameter Λ is 10^3.) This means that the time between dissociation events is much longer than the time to diffuse a distance L, so the process is limited by the speed of the reactions, not by the diffusion of the rotor. In other words, the diffusive motion of the rotor is so fast that it achieves thermodynamic equilibrium, and so its displacement can be described by a Boltzmann distribution. In this case we can express the probability distribution as $p_p(x, t) = p_p(t)\cdot P(x)$, where $p_p(t)$ is the probability of the site being in the protonated state and $P(x)$ is the *equilibrium* spatial probability density describing the rotor's position relative to the stator.

We can obtain the steady-state Boltzmann distribution $P(x)$ from (17). First, we divide through by Λ and take advantage of the fact that $\Lambda >> 1$: All terms but the last are rendered negligible, so that the distribution of rotor positions in the protonated state potential well ϕ_p $(0 \leq x \leq 1)$ is governed by

$$\frac{dp}{dx} - wP = 0.$$

The solution must be normalized to 1, since it represents a probability density. The result is:

$$P(x) = \left(\frac{w}{e^w - 1}\right)e^{wx}, \; 0 \leq x \leq 1, \qquad ...(19)$$

where the quantity in parentheses is the normalization factor. Thus, the rates of protonation at time t are $p_p(0,t) = p_p(t)\cdot P(0)$ and $p_p(l, t) = p_p(t).P(1)$, respectively. Substituting this into (15) and imposing the conservation of probability, $p_d(t) + p_p(t) = 1$, we reduce the problem to the two-state Markov chain described by:

$$\frac{dp_d}{dt} = -\frac{dp_p}{dt} = -kp_d + (P_+ + P_-)p_p, \qquad ...(20)$$

where $P_+(w) = w/(1 - e^{-w})$, $P_-(w) = w/(e^w - 1)$. Therefore, the stationary probabilities are obtained directly by setting the time derivatives equal to zero and solving for p_p:

$$p_d(w,k) = \frac{P_+ + P_-}{k + P_+ + P_-}, \quad p_p(w,k) = \frac{k}{k + P_+ + P_-}, \qquad ...(21)$$

or, in the dimensional variables:

$$P_d(w,k) = \frac{k_d P_+ + k_d P_-}{k_p(c^{\text{acid}} + c^{\text{base}})L + k_d P_+ + k_d P_-}, \qquad ...(22)$$

$$P_p(w,k) = \frac{k_p(c^{\text{acid}} + c^{\text{base}})L}{k_p(c^{\text{acid}} + c^{\text{base}})L + k_d P_+ + k_d P_-}, \qquad ...(23)$$

The average velocity of the motor can be found using the following heuristic argument. The rotor effectively moves to the right either from the protonated state, when the proton is released to the basic reservoir with the effective rate $k_d P_+$, or from the deprotonated state, when the protonation takes place at the acidic reservoir with the effective rate $k_p c^{\text{acid}} L$. The corresponding effective rate of movement to the right is the sum of the corresponding rates weighted by the respective state probabilities: $\langle V_r \rangle = k_p c^{acid} L p_d + k_d P_+ p_p$. Similarly, the rotor effectively moves to the left either from the protonated state, when the proton is released to the acidic reservoir with the effective rate $k_d P_-$, or from the deprotonated state, when the protonation takes place at the basic reservoir with the effective rate $k_p c^{\text{base}} L$. The corresponding effective rate of movement to the left is the sum of the corresponding rates weighted by the respective state probabilities: $\langle V_l \rangle = k_p c^{\text{base}} L p_d + k_d P_- p_p$. The net average velocity $\langle V \rangle = \langle V_r \rangle - \langle V_l \rangle$, can be obtained by using the expressions for the state probabilities (22), (23) and some algebra:

$$\langle V \rangle (F_L) = \frac{k_p k_d L w (c^{\text{acid}} e^w - c^{\text{base}})}{(e^w - 1)k_p L(c^{\text{acid}} + c^{\text{base}}) + k_d w(e^w + 1)},$$

$$w = \frac{(F_I - F_L)L}{k_B T} \qquad ...(24)$$

As a check, note that when there is no load ($F_L = 0$), no membrane potential ($F_I = 0$), and no proton gradient ($c^{\text{acid}} = c^{\text{base}}$), the velocity vanishes, as it should. The load-velocity relationship given by (24) looks very similar to the one in the next limiting case.

The *stall force* F_s is reached when the load force just brings the motor to a halt ($c^{acid} e^{w} - c^{base} = 0$):

$$F_s = F_l + \frac{k_B T}{L} \ln\left[\frac{c^{\text{acid}}}{c^{\text{base}}}\right] \quad ...(25)$$

Since the electrical driving force satisfies

$$F_l = e\Delta\psi - 2.3 k_B T \Delta\text{pH} \quad ...(26)$$

This says that the reversible work done to move a rotor site across the membrane is equal to the work done by the electrical field plus the "entropic work" done by the Brownian ratchet. (The term "reversible" in this context means that the velocity is so slow (near stall) that we can neglect the viscous dissipation.) Dividing through by the unit charge e gives the work per unit charge, which is just the protonmotive force discussed previously. One point about (26) is worth noting. Since the motor is working against a conservative load force, as the motor approaches stall its efficiency approaches 100%. For a motor working against a viscous load, a more sophisticated treatment is required.

It may seem from (24) that there is a definite average velocity of the motor in the limit $T \rightarrow \infty$. In other words, the motor continues to move in the absence of thermal fluctuation. The reason is that the solution of the Lengevin and Smoluchowski equations cannot be treated as a regular perturbation problem in the limit of low temperatures. This is a singlular perturbation problem, and the protein behavior at absolute zero temperature cannot be quantified as a simple limit of such behavior at low temperatures.

Brownian Ratchet vs. Power Stroke

In the last example the motion of the rotor was driven by a combination of Brownian motion and the membrane potential. The rotor diffusion is biased by the electrostatic forces that are switched off and on by the binding and dissociation of protons to the rotor sites. The membrane potential appears to drive the rotor unidirectionally without the aid of Brownian motion. However, this cannot happen without the binding and dissociation of protons, a stochastic process driven by thermally excited transitions. Thus even the "power stroke" depends on

Brownian motion, so that setting $k_B T = 0$ in the model equations arrests the rotor motion. This is a fundamental distinction between molecular and macroscopic motors. The distinction between a motor driven by a "Brownian ratchet" and one driven by a "power stroke" may not be so clear in other systems.

In the polymerization ratchet model described above, the movement of the *load* is driven entirely by its Brownian motion. The chemical step simply rectifies, or biases, this motion. By comparison, the FI motor in ATP synthase is driven by the hydrolysis of ATP. The conformational change in the protein that constitutes the power stroke is known: Binding of ATP to the catalytic site drives the change in protein shape that drives rotation. That is, the load is not driven significantly by Brownian motion; it sees the proteins conformational change as a "power stroke."

However, a closer look at how ATP binds to the catalytic site reveals that it is a multistep process involving the sequential annealing of hydrogen bonds between the protein and the nucleotide. Each step in this process is driven by Brownian motion, i.e., a thermally activated process. Therefore, the power stroke itself can be viewed as a kind of Brownian ratchet, one that proceeds at a smaller length scale than the protein (ATP synthase is % 10 nm in diameter, while the catalytic site is * 1 nm). Thus the distinction between a process driven by a Brownian ratchet and by a power stroke can be largely a matter of size scale; a fuzzy boundary separates the two notions. In the extreme case where the motion of the *load* is due only to its diffusion, and the role of the chemical reaction is only to inhibit diffusion in one direction, we can say that the motor is a Brownian ratchet.

Average Velocity of the Motor when Chemical Reactions are as Fast as Diffusion

Next we consider a different limiting case: where the diffusion time scale is comparable to the reaction rates in the basic reservoir. In this case we have to change the mathematical formulation of the model. We make the simplifying assumption that the proton concentration in the acidic reservoir is so high that the binding sites on that side are always protonated. Now we define the right boundary of the membrane as the origin, and the distance between the membrane and the binding site nearest the membrane in the basic reservoir x is always between 0 and L. The chemical state of the rotor is determined by the state of all the binding sites in the acidic reservoir. In general, if there are N binding sites on this side, the total number of chemical states is 2^N. However, for now we focus on the binding site nearest the membrane. In this case there are just two states: "off" if this site is unprotonated and "on" if it is protonated. In previous discussion, the mechanochemistry of the motor was described by the following set of coupled diffusion equations:

$$\frac{\partial p_d}{\partial t} = D\frac{\partial}{\partial x}\left(\frac{F_I - F_I}{k_B T}p_d + \frac{\partial p_d}{\partial x}\right) - k_p p_d + k_d p_p, \quad \text{...(27)}$$

$$\frac{\partial p_d}{\partial t} = D\frac{\partial}{\partial x}\left(\frac{F_L - F_I}{k_B T}p_p + \frac{\partial p_p}{\partial x}\right) + k_p p_d - k_d p_p, \quad \text{...(28)}$$

where $p_p(x, t)$ and $P_d(x,t)$ (1/nm) are the probability densities for being at position x and in the protonated and deprotonated states, respectively, at time t. The proton association and dissociation rates in basic reservoir are k_p (1/s) *and* k_d (1/s), respectively. Note that dimensions of some of the model parameters and variables are different in this limit.

We can nondimensionalize these equations using the rescaled coordinate $(xlL) \rightarrow x$, the re-scaled time $k_d t \rightarrow t$, and the dimensionless parameters $\Lambda = (D/k_d L^2)$, $w = (F_I - F_L)\, L/\kappa_B T$, and $k = k_p/k_d$. The nondimensional equations have the form

$$\frac{\partial p_d}{\partial t} = \Lambda\frac{\partial}{\partial x}\left(wp_d + \frac{\partial p_d}{\partial x}\right) - kp_d + p_p, \quad \text{...(29)}$$

$$\frac{\partial p_p}{\partial t} = \Lambda\frac{\partial}{\partial x}\left(wp_d + \frac{\partial p_p}{\partial x}\right) + kp_d - p_p. \quad \text{...(30)}$$

Equations (29) and (30) are second-order partial differential equations. This means that four boundary conditions are required in order to have a mathematically complete description of the problem. One boundary condition is that* = 0 is reflecting:

$$\left[wp_d + \frac{\partial p_d}{\partial x}\right]_{x=0} = 0 \quad \text{...(31)}$$

This takes into account the fact that an unprotonated site cannot pass back through the membrane. The remaining three boundary condtions require knowing the state of all the binding sites in the basic reservoir, which would necessitate solving a large number of coupled diffusion equations (one for each possible chemical state of the rotor). However, to keep things simple, we construct a reflecting boundary condition at $x = 1$ (x is now measured in units of L) when the rotor is in the protonated state. That is,

$$\left[wp_p + \frac{\partial p_p}{\partial x}\right]_{x=1} = 0 \qquad \text{...(32)}$$

If the proton dissociation rate is fast and the proton concentration is low in the basic reservoir, we don't expect this artificial boundary condition to have much of an effect, since in this limit the probability of the first site being occupied when $x = I$ is very small, so that this boundary condition is rarely encountered.

When an unprotonated site moves to the right of $x = 1$, it brings a protonated site out of the membrane channel and into the region $0 < x < 1$. This protonated site becomes the new site that we follow. The state of the motor goes from unprotonated to protonated, and the coordinate of the motor goes from $x = I$ to $x = 0$. Conversely, when a protonated site moves into the membrane, it brings an unprotonated site into the region $0 < x < 1$. This unprotonated site becomes the new site we follow. The state and the coordinate of the motor change accordingly. These considerations are illustrated in Figure 1.6 by the arrows connecting the right end of p_d to the left end of p_p. The boundary conditions that model this situation are:

$$\left(wp_d + \frac{\partial p_d}{\partial x}\right)_{x=1} = \left(wp_p + \frac{\partial p_p}{\partial x}\right)_{x=0}, \quad p_d(1,t) = p_p(0,t), \qquad \text{...(33)}$$

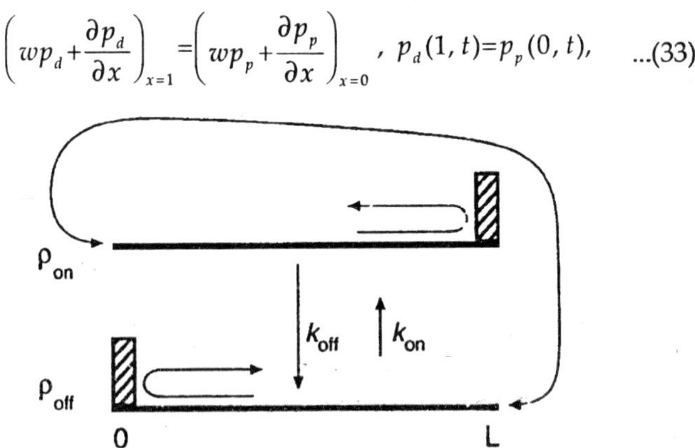

Fig. 1.6. In general, the chemical state of the motor is determined by all the binding sites in the right chamber. In the simplified version of the model only the binding site nearest the membrane is considered.

which is the mathematical statement of the fact that the rotor in the off state at $x = 1$ is equivalent to the rotor in the on state at $x = 0$. Therefore, to implement these boundary conditions numerically, we make use of the periodic boundary condition discussed in the previous chapter. Note that there are two mechanisms for changing the chemical state of the rotor: movement of the rotor and chemical kinetics.

We are now in a position to use the numerical algorithm to approximate (29)-(33). The interval (0,1) is divided into M segments. For each of the M grid points there are two possible states of the rotor, off and on. Therefore, there are 2M possible states in the discrete approximation of the process. Let the first M states correspond to the offstate, and the states M +1 to $2M$ correspond to the onstate. The equations used in the numerical scheme for $1 < n < M$ *are*

$$\frac{dp_n}{dt}=(F_{n-1/2p_{n-1}}-B_{n-1/2p_n})-(F_{n+1/2p_n}-B_{n+1/2p_{n+1}})-kp_n+p_n+M,$$

$$\frac{dp_n+M}{dt}=(F_{n-1/2p_n-1+M}-B_{n-1/2p_{n+M}})-(F_{n+1/2p_{n+M}}-B_{n+1/2p_n+1+M})-p_n+M+kp_n$$

and the equations used to implement the boundary conditions are

$$\frac{dp_1}{dt} = -(F_{3/2p_1}-B_{3/2}p_2)-kp_1+p_{M+1}, \quad ...(35)$$

$$\frac{dpM}{dt} = (F_{M-1/2p_{M-1}}-B_{M-1/2pM})-(F_{1/2pM}-B_{1/2pM+1})-kp_M+p_{2M}, \quad ...(36)$$

$$\frac{dp_{M+1}}{dt} = (F_{1/2p_M}-B_{1/2p_{M+1}})-(F_{3/2p_{M+1}}-B_{3/2p_{M+2}})-p_{M+1}+kp_1, \quad ...(37)$$

$$\frac{dp_{2M}}{dt} = (F_{M-1/2p2M-1}-B_{M-1/2p2M})-p_{2M}+k_{pM} \quad ...(38)$$

The potential used in F_n and B_n is $\phi(x) = (F_L - F_I)\,x$. Note that because of the chemical kinetics, the matrices A and C required for the numerical scheme are no longer tridi-agonal. However, they are still sparse, so that solving (39) is not computationally expensive.

Let us discuss briefly how to calculate the protonation and deprotonation rates k_p and k_d. The chemical reaction is

$$\text{site}^- + H^+ \leftrightarrow \text{site. H.} \quad ...(39)$$

At equilibrium, protonation and deprotonation balance. That is,

$$k_p[\text{site}^-] = k_d[\text{site} \cdot \text{H}]. \qquad ...(40)$$

Proton concentrations are generally reported as a pH value:

$$\text{pH} = -\log_{10}[\text{H+}], \quad [\text{H+}] = 10^{-\text{pH}} \qquad ...(41)$$

The higher the pH value, the lower the proton concentration. The pK_a value of the binding site is calculated from the measured concentration values of [site · H] and [site⁻] as

$$\text{pK}_a = \text{pH} + \log_{10}\frac{[\text{site}\cdot\text{H}]}{[\text{site}^{-1}]} \qquad ...(42)$$

Combining (40) and (42), we see that the rates k_p and k_d are related to pH and pK_a by

$$\frac{\bar{k}_p}{k_d} = 10^{pK_a - pH} \qquad ...(43)$$

Generally, k_p is limited by the rate at which protons diffuse to the binding site. In this limit, the association rate can be computed from the Smoluchowski formula (this rate is proportional to the proton concentration, c^{base}):

$$\bar{k}_p = \begin{pmatrix} \text{proton} \\ \text{concentration} \end{pmatrix} \cdot \begin{pmatrix} \text{absorption rate to a perfectly} \\ \text{absorbing disk of radius } r \end{pmatrix}$$

$$= \underbrace{0.6\ \text{nm}^{-3} \cdot 10^{-\text{pH}}}_{\text{protons/nm}^3} \quad \underbrace{4r\,D_{\text{proton}}}_{\text{absorption rate}} \qquad ...(44)$$

Here D_{proton} is the diffusion coefficient of protons. If we know k_p, k_d can be determined by (43). The table shown below lists typical parameter values for ATP synthase.

Parameter name	*Parameter Value*
Diffusion coefficient of the rod	$D = 10^4$ nm²/s
pK_a value of binding site	$\text{pK}_a = 6.0$
pH of the right compartment	pH = 6.0to 8.0 (variable)
External loadforce on the rod	$F_L = 0$ to 3 pN (variable)
Distance between binding sites	$L = 8$ nm
Diffusion coefficient of proton	$D_{\text{proton}} = 10^{10}\text{nm}^2/\text{s}$
Absorbing radius of binding site	$r = 0.5$ nm

Note that at these values of the model parameters, at pH = 7, the values of the nondimensional parameters are $\Lambda \approx 0.1$, $k \approx 0.1$, $w \approx 1$.

We now have all the necessary information to use the numerical scheme. Figure 1.7A shows the relaxation of the

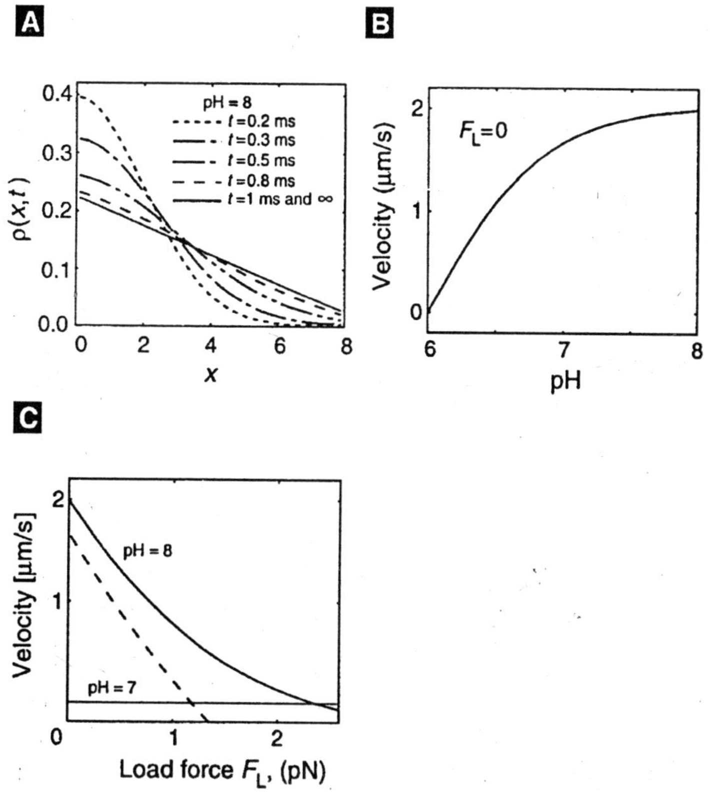

Fig. 1.7. Numerical results of the simplified F_0 motor.

marginal density $p(x, t) = p_p(x, t\} + p_d(x, t)$ to steady state. The (dimensional) average velocity is a steady-state property of the system and is related to the total flux by the relations

$$\langle V \rangle = LJ = -LD\left[\frac{F_L - F_I}{k_B T}\left(p_p^{(s)} + p_d^{(s)}\right) + \frac{\partial}{\partial x}\left(p_p^{(s)} + p_d^{(s)}\right)\right], \quad ...(45)$$

where the superscript 5 in the above equation stands for steady state. Once steady state is achieved, the average velocity *(V)* can then be computed from the relation

$$\langle V \rangle = L(F_{n-1/2pn-1} - B_{n-1/2p_n}) + L(F_{n-1/2p_{n+M}}) \quad ...(46)$$

for $1 < n < M$.

Other Motor Proteins

There is such a variety of protein motors that no classification scheme can do justice to their diversity. However, for the purposes of discussion we can identify several physical properties that delineate classes of motors. Some of the most important characteristics of molecular motors are the following:

Fuel

The two most common energy sources for molecular motors are nucleotide hydrolysis (e.g., ATP, GTP) and transmembrane ion gradients. Certain specialized motors depend on stored elastic energy that has been captured during the assembly of the motor (e.g., spasmoneme and the acrosome of *Limulus*).

Mechanical Escapement

Three common motor types are (i) rotary (e.g., the bacterial flagellar motor FQ ATPase); (ii) linear motors that run along a "track," usually actin, microtubules, or nucleic acid polymers (e.g., myosin, kinesin, dynein, RNA polymerase); (iii) polymerization or depolymerization motors that directly push or pull against a load (e.g., the acrosome, cellular lamellipodia, the propulsive tail of *Listeria*). The latter category suggests a subclassification into those that operate in a continuous cycle (e.g., myosin, FI), and motors that are "one-shot" processes; i.e., they function for but a single episode of polymerization, after which they are usually disassembled.

Cooperative vs. "Loners"

Because of its small "duty cycle" (i.e., attachment time to the load), myosin II must act in concert with many other partners to produce a continuous force on an actin filament. Myosin V and kinesin, however, have longer duty cycles, and so they can transport a vesicle without the cooperation of other motors.

These categories do not begin to delineate the variety of possibilities. However, one common event generally commences the transduction process between chemical energy and mechanical force. Because molecular motors can be viewed as enzymes, the binding of a substrate onto the motor initiates the transduction process. However, this does not really tell us much, since it is simply a restatement that chemical reactions (other than isomerizations) begin by combining substrates. The feature that distinguishes molecular motors from other enzyme reactions is that the binding event is directly or indirectly coupled to the creation of mechanical forces. For example, in FI ATPase, the binding of ATP to the catalytic site directly generates the power stroke. However, the coupling may not always be so direct. For example, binding of a proton to the FQ rotor site switches off the local electrostatic field surrounding the rotor, permitting bidirectional diffusion. The binding of monomers to polymerizing actin captures thermal fluctuations in elastic strain, which is subsequently released to power protrusion.

Several points are important. First, the role of thermal fluctuations in all these processes is central, so that none could operate when $k_B T = 0$. Second, energy captured by binding or dissociation events can be stored and released later, and in a different location, to produce mechanical work. Third, the operation of every molecular motor depends on its specialized protein geometry, so that models of motors that ignore geometry are generally not useful to biologists.

Finally, we do not believe that the operation of molecular motors involves any novel physics or chemistry. However, the amazing variety of protein shapes requires that we treat each motor individually. General principles are not likely to provide more than philosophical comfort in understanding any particular motor.

2

COMPUTATIONAL DIVERGENCE

SEQUENCE ALIGNMENT

Given two or more sequences, we initially wish to

- measure their similarity
- determine the residue-residue correspondences
- observe patterns of conservation and variability
- infer evolutionary relationships

If we can do this, we will be in a good position to go fishing in databanks for related sequences. A major application is to the annotation of genomes, involving assignment of structure and function to as many genes as possible.

How can we define a quantitative measure of sequence similarity? To compare the nucleotides or amino acids that appear at corresponding positions in two or more sequences, we must first assign those correspondences. *Sequence alignment is the identification of residue-residue correspondences.* It is *the* basic tool of bioinformatics.

Any assignment of correspondences that preserves the *order* of the residues within the sequences is an alignment. Gaps may be introduced.

Given two text strings:

first string = a b c d e   second string = a c d e f

a reasonable alignment would be

```
a b c d e –
a – c d e f
```

We must define criteria so that an algorithm can choose the *best* alignment. For the sequences gctgaacg and ctataatc:

An uninformative alignment	– – – – – – – g c t g a a c g
	c t a t a a t c – – – – – – –
An alignment without gaps	g c t g a a c g
	c t a t a a t c
An alignment with gaps	g c t g a – a – – c g
	– – c t – a t a a t c
And another	g c t g – a a – c g
	– c t a t a a t c –

Most readers would consider the last of these alignments the best of the four. To decide whether it is the best of *all* possibilities, we need a way to examine all possible alignments systematically. Then we need to compute a score reflecting the quality of each possible alignment, and to identify the alignment with the optimal score. The optimal alignment may not be unique: several different alignments may give the same best score. Moreover, even minor variations in the scoring scheme may change the ranking of alignments, causing a different one to emerge as the best.

The Dotplot

The *dotplot* is a simple picture that gives an overview of the similarities between two sequences. Less obvious is its close relationship to alignments.

The dotplot is a table or matrix. The rows correspond to the residues of one sequence and the columns to the residues of the other sequence. In its simplest form, the positions in the dotplot are left blank if the residues are different, and filled if they match. Stretches of similar residues show up as diagonals in the upper left-lower right (Northwest-Southeast) direction.

The dotplot gives a quick pictorial statement of the relationship between two sequences. Obvious features of similarity stand out. For example, a dotplot relating the mitochondrial ATPase-6 genes from a lamprey (*Petromyzon marinus*) and dogfish shark (*Scyliorhinus canicula*) shows that the similarity of the sequences is weakest near the beginning. This gene codes for a subunit of the ATPase complex. In the human, mutations in this gene cause Leigh syndrome, a neurological disorder of infants produced by the effects of impaired oxidative metabolism on the brain during development. A disadvantage of the dotplot is that its 'reach' into the realm of distantly-related sequences is poor. In analysing sequences, one should always look at a dotplot to be sure of not missing anything obvious, but be prepared to apply more subtle tools.

Often regions of similarity may be displaced, to appear on parallel but not collinear diagonals. This indicates that insertions or deletions have occurred in the segments between the similar regions. A dotplot relating the PAX-6 protein of mouse and the eyeless protein of *Drosophila melanogaster* shows three extended regions of similarity with different lengths of sequence between them, two near the beginning of the sequences and one near the middle. Between the second and third of them, there is a longer intervening region in the mouse than in the *Drosophila* sequence.

Filtering the results can reduce the noise in a dotplot. In the comparison of the ATPase sequences, dots were not shown unless they were at the centre of a consecutive region of 15 residues containing at least 6 matches. The PERL program for dotplots allows the user to set values for a window (length of region of consecutive residues) and a *threshold* (number of matches required within the window).

Dotplots and Sequence Alignments

The dotplot captures in a single picture not only the overall similarity of two sequences, but also the complete set and relative quality of different possible alignments. Any path through the dotplot from upper left to lower right, moving at each point only East, South or Southeast, corresponds to a possible alignment. If two sequences are closely related, the alignment can be read directly off the dotplot.

If the direction of the 'move' between successive cells is diagonal, two pairs of successive residues appear in the alignment without an insertion between them. If the direction of the move is horizontal, a gap is introduced in the sequence indexing the rows. If the direction of the move is vertical, a gap is introduced in the sequence indexing the columns. Note that no moves can be directed up or to the left, as this would correspond to aligning several residues of one sequence with only one residue of the other. The path indicated by the arrows corresponds to the obvious alignment:

```
DOROTHY--------HODGKIN
DOROTHYCROWFOOTHODGKIN
```

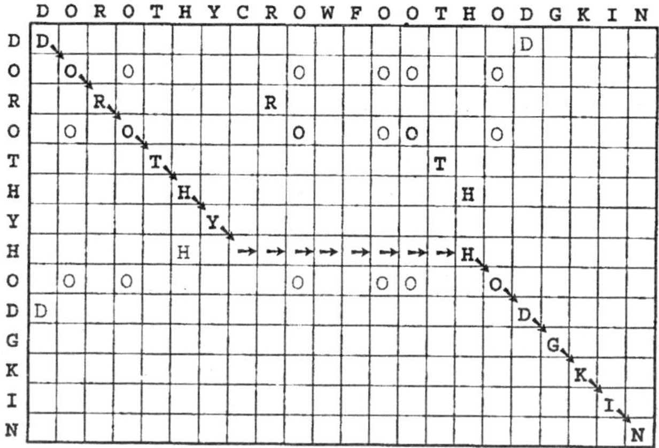

Fig. 2.1. Any path through the dotplot from upper left to lower right passes through a succession of cells, each of which picks out a pair of positions, one from the row and one from the column, that correspond in the alignment; or that indicates a gap in one of the sequences.

Another way to think of a path through the dotplot is as an *edit script*; that is, the prescription of a series of operations that transforms the sequence that indexes the columns—the 'horizontal' sequence—into the sequence that indexes the rows—the 'vertical' sequence. Each move tells us to perform an operation—a substitution, an insertion, or a deletion. When the end of the path is reached, the effect will be to change one sequence into the other. In general, several different sequences of edit operations may convert one string into the other in the same number of steps, but they may induce different alignments.

It should be emphasized that although a sequence of edit operations derived from an optimal alignment *may* correspond to an actual evolutionary pathway, it is impossible to *prove* that it does. The larger the edit distance, the larger the number of reasonable evolutionary pathways between two sequences.

Measures of Sequence Similarity

To go beyond 'alignment by eyeball' via dotplots, we must define quantitative measures of sequence similarity and difference. Given two character strings, two measures of the distance between them are:

1. The *Hamming* distance, defined between two strings of equal length, is the number of positions with mismatching characters.

2. The *Levenshtein,* or edit distance, between two strings of not necessarily equal length, is the minimal number of 'edit operations' required to change one string into the other, where an edit operation is a deletion, insertion or alteration of a single character in either sequence. A given sequence of edit operations induces a unique alignment, but not *vice versa.*

   For example:

agtc cgta	Hamming distance = 2
ag–tcc cgctca	Levenshtein distance = 3

For applications to molecular biology, recognize that certain changes are more likely to occur naturally than others. For example, amino acid substitutions tend to be conservative: the replacement of one amino acid by another with similar size or physicochemical properties is more likely to have occurred than its replacement by another amino acid with greater difference in their properties. Or, the deletion of a succession of contiguous bases or amino acids is a more probable event than the independent deletion of the same number of bases or amino acids at non-contiguous positions in the sequences. Therefore, we may wish to assign variable weights to different edit operations. A computer program can then determine not just minimal edit distances but optimal alignments. It can score each path by adding up the scores of the individual steps. For substitutions, it adds the score of the mutation, depending on the pair of residues involved. For horizontal and vertical moves, it adds a suitable gap penalty.

Scoring Schemes

A scoring system must account for residue substitutions, and insertions or deletions. Deletions, or gaps in a sequence, will have scores that depend on their lengths.

Hamming and Levenshtein distances measure the *dissimilarity* of two sequences: similar sequences give small distances and dissimilar sequences give large distances. It is common in molecular biology to define scores as measures of sequence *similarity*. Then similar sequences give high scores and dissimilar sequences give low scores. These are equivalent formulations. Algorithms for optimal alignment can seek either to minimize a dissimilarity measure, or maximize a scoring function.

For nucleic acid sequences, it is common to use a simple scheme for substitutions: +1 for a match, –1 for a mismatch, or a more complicated scheme based on the higher frequency of transition mutations than transversion mutations.

For proteins, a variety of scoring schemes have been proposed. We might group the amino acids into classes of similar physicochemical type, and score +1 for a match within residue class, and –1 for residues in different classes. We might try to devise a more precise substitution score from a combination of properties of the amino acids. Alternatively, we might try to let the proteins teach us an appropriate scoring scheme. M.O. Dayhoff did this first, by collecting statistics on substitution frequencies in the protein sequences then known. Her results were used for many years to score alignments. They have been superseded by newer matrices based on the very much larger set of sequences that has subsequently become available.

Derivation of substitution matrices

As sequences diverge, mutations accumulate. To measure the relative probability of any particular substitution, for instance Serine→Threonine, we can count the number of Serine → Threonine changes in pairs of aligned homologous sequences. We could use the relative frequencies of such changes to form a scoring matrix for substitutions. A likely change should score higher than a rare one. But, what if there have been multiple substitutions at certain sites? This will bias the statistics. We can avoid this problem by restricting our samples to sequences that are sufficiently similar so that we can assume that no position has changed more than once.

A measure of sequence divergence is the PAM: 1 PAM = 1 Percent Accepted Mutation. Thus, two sequences 1 PAM apart have 99% identical residues. For pairs of sequences within the 1 PAM level of divergence, it is likely that there has been no more than one change at any position. Collecting statistics from pairs of sequences as closely related as this, and correcting for different amino acid abundances, produces the *1 PAM substitution matrix*. To produce a matrix appropriate for more widely divergent sequences, we can take powers of this matrix. The PAM250 level, corresponding to ~20% overall sequence identity, is the lowest sequence similarity for which we can hope to produce a correct alignment by sequence analysis alone. It is therefore the appropriate level to choose for practical work. (Several authors have derived substitution matrices appropriate in different ranges of overall sequence similarity.)

The occurrence of reversions, either directly or via one or more other changes, produces an apparent slowdown in mutation rates as sequences progressively diverge. The relation between PAM score and % sequence identity is:

PAM	0	30	80	110	200	250
% identity	100	75	50	60	25	20

It expresses scores as *log-odds* values:

Score of mutation $i \leftrightarrow j$

$$= \log \frac{\text{observed } i \leftrightarrow j \text{ mutation rate}}{\text{mutation rate expected from amino acid frequencies}}$$

The numbers are multiplied by 10, simply to avoid decimal points. The matrix entries reflect the probabilities of a mutational event. A value of +2 (for instance, C $\leftrightarrow$ S) implies that in related sequences the mutation would be expected to occur 1.6 times more frequently than random. The calculation is as follows: The matrix entry 2 corresponds to the actual value 0.2 because of the scaling. The value 0.2 is $\log_{10}$ of the relative expectation value of the mutation. Therefore, this expectation value is $10^{0.2} = 1.6$.

The probability of two independent mutational events is the product of their probabilities. By using logs, we have scores that we can add up rather than multiply, a computational convenience.

The BLOSUM matrices

S. Henikoff and J.G. Henikoff developed the family of BLOSUM matrices for scoring substitutions in amino acid sequence comparisons. Their goal was to replace the Dayhoff matrix with one that would perform best in identifying distant relationships, making use of the much larger amount of data that had become available since Dayhoff's work.

The BLOSUM matrices are based on the BLOCKS database of aligned pro, tein sequences; hence the name BLOcks SUbstitution Matrix. From regions of closely-related proteins alignable without gaps, Henikoff and Henikoff calculated the ratio, of the number of observed pairs of amino acids at any position, to the number of pairs expected from the overall amino acid frequencies. As in the Dayhoff matrix, the results are expressed as log-odds.

In order to avoid over-weighting closely-related sequences, the Henikoffs replaced groups of proteins that have sequence identities higher than a threshold by either a single representative or a weighted average. The threshold 62% produces the commonly used BLOSUM62 substitution matrix. This is offered by all programs as an option and is the default in most. BLOSUM matrices have largely replaced the Dayhoff matrix for most applications.

Scoring insertions/deletions, or 'gap weighting'

To form a complete scoring scheme for alignments, we need, in addition to the substitution matrix, a way of scoring gaps. How important are insertions and deletions, relative to substitutions? Distinguish gap initiation

```
aaagaaa
aaa–aaa
```

from gap extension:

```
aaaggggaaa
aaa- - - -aaa
```

For aligning DNA sequences, CLUSTAL-W recommends use of the identity matrix for substitution (+1 for a match, 0 for a mismatch) and gap penalties 10 for gap initiation and 0.1 for gap extension by one residue. For aligning protein sequences, the recommendations are to use the BLOSUM62 matrix for substitutions, and gap penalties 11 for gap initiation and 1 for gap extension by one residue.

Computing the Alignment of Two Sequences

Now that we have a scoring scheme, we can apply it to finding optimal alignments—we seek the alignment that maximizes the score. A famous algorithm to determine the global optimal alignments of two sequences is based on a mathematical technique called dynamic programming.This algorithm has been extremely important in molecular biology. Two of several noteworthy features are

- The good news is that the method is guaranteed to give a *global* optimum. It will find the *best* alignment score, given the choice of parameters—substitution matrix and gap penalty—with no approximation.
- The bad news is that many alignments may give the same optimal score. And none of these need correspond to the

biologically correct alignment. For instance, in comparing the α- and β-chains of chicken haemoglobin, W. Fitch and T. Smith found 17 alignments all of which give the same optimal score, one of which is correct (on the basis of the structures, the court of last resort). There are 1317 alignments with scores within 5% of the optimum.

Another item of bad news is technical: The time required to align two sequences of lengths n and m is proportional to $n \times m$, because this is the size of the edit matrix that must be filled in. This means that the dynamic-programming method is not convenient to use for searching in an entire sequence database for a match to a probe sequence, and even less convenient for 'all-against-all' alignments. The database search problem is in effect the problem of matching a probe sequence to a region of a very long sequence, the length of the entire database.

Variations and Generalizations

Variations of this method apply to three related alignment questions: entire sequence against entire sequence, region of one sequence against entire other sequence, and region of one sequence against region of other sequence. The global alignment algorithm was first applied to biological sequence alignment by S.B. Needleman and C.D. Wunsch. T. Smith and M. Waterman modified it to identify local matches.

Approximate Methods for Quick Screening of Databases

It is routine to screen genes from a new genome against the databases for similarity to other sequences. Approximate methods can detect close relationships well and quickly but are inferior to the exact ones in picking up very distant relationships. In practice, they give satisfactory performance in the many cases in which the probe sequence is fairly similar to one or more sequences in the databank, and they are therefore worth trying first.

A typical approximation approach would take a small integer k, and determine all instances of each k-tuple of residues in the probe sequence that occur in any sequence in the database. A candidate sequence is a sequence in the databank containing a large number of matching k-tuples, with equivalent spacing in probe and candidate sequences. For a selected set of candidate sequences, approximate optimal alignment calculations are then carried out, with the time- and space-saving restriction that the paths through the matrix considered are restricted to bands around the diagonals containing the many matching *k*-tuples. There are several variations on this theme.

Dynamic Programming Algorithm for Optimal Pairwise Sequence Alignment

A chart implicitly containing all possible alignments can be constructed as a matrix similar to that used in drawing the dotplot. The residues of one sequence index the rows, the residues from the other sequence index the columns. Any path through the matrix from upper left to lower right corresponds to an alignment. The task is to find the path that has the lowest cost, and the difficulty is that there are a very large number of paths to consider.

As an illustration, suppose you wanted to drive from Malmö in southern Sweden to Tromsø in Northern Norway. Your route will consist of a number of segments, taking you through a succession of intermediate cities. There are many choices of different combinations of segments to produce a complete, continuous path.

The computational approach to finding the optimal path begins by assigning a numerical measure of the 'cost' to each of the possible individual segments of the journey. This 'cost' is not simply the financial outlay, but a more general estimate of your relative preferences for different portions of the route. The distance travelled will clearly be an important component of the cost, but other factors such as the quality of the roads and the opportunities for sightseeing also contribute. For any route selected, the overall cost of the trip is the sum of the costs of the individual segments. Clearly, it is inefficient to repeat any leg of the journey, or to visit any city twice, so we shall agree that every intermediate stop will be North of the previous one. This formalism is expressed in terms of minimizing a cost rather than maximizing a score; for our purposes the two approaches are equivalent. An algorithm can explore the possible combinations to determine an optimal overall route.

Here is an abstract version of the problem: to grasp the essential idea of dynamic programming, first consider: how many paths from Start to Finish pass through A? There are 6 paths from Start to A. (Write them all down.) Therefore, by symmetry there

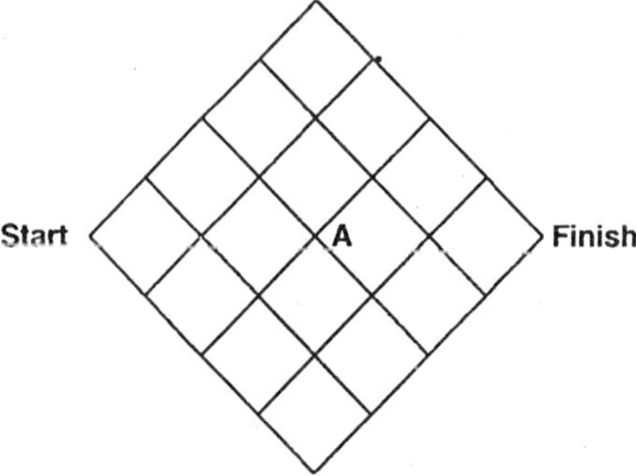

are 6 paths from A to finish, and a total of 36 paths from Start to Finish passing through A. (Why?) Assuming that we have assigned costs to the individual steps, do we have to check all 36 paths to find the path of minimum cost that goes from Start to Finish, passing through A? No—here is the crucial observation: *The choice of the best path from A to Finish is independent of the choice of path from the Start to A.* If we determine the best of the 6 paths from Start to A, and we determine the best of the 6 paths from A to Finish, the best path from Start to Finish passing through A is: the best path from Start to A *followed* by the best path from A to finish. No more than 12 of the paths through A need be considered.

Even greater simplification is possible by systematically resubdividing the problem. The dynamic programming method for finding the optimal path through the matrix is based on this idea.

A statement of the optimal alignment problem and the dynamical programming solution are as follows: Given two character strings, possibly of unequal length: $A = a_1a_2 \dots a_n$ and $B = b_1b_2 \dots b_m$, where each a_i and b_j is a member of an alphabet set A, consider sequences of edit operations that convert A and B to a common sequence. Individual edit-operations include:

Substitution of b_j for a_i—represented (a_i, b_j)

Deletion of a_i from sequence A—represented (a_i, ϕ).

Deletion of b_j from sequence B—represented as (ϕ, b_j).

If we extend the alphabet set to include the null character ϕ: $A^+ = A \cup \{\phi\}$, a sequence of edit operations is a set of ordered pairs (x, y), with $x, y \in A^+$.

A *cost function* d is defined on edit operations: $d(a_i, b_j)$ = cost of a mutation in an alignment in which position i of sequence A corresponds to position j of sequence B, and the mutation substitutes $a_i \leftrightarrow b_j$. $d(a_i, \phi)$ or $d(\phi, b_j)$ = cost of a deletion or insertion. Define the minimum weighted distance between sequences A and B as

$$D(A,B) = \min_{\to B} \sum d(x, y)$$

where $x, y \in A^+$ and the minimum is taken over all sequences of edit operations that convert A and B into a common sequence.

The problem is to find $D(A, B)$ and one or more of the alignments that correspond to it.

An ,algorithm that solves this problem in $O(mn)$ time creates a matrix $D(i,j)$, $i = 0,\ldots n; j = 0,\ldots m$, such that $D(i,j)$ is the minimal distance between the strings that consist of the first i characters of A and the first j characters of B. Then $D(n, m)$ will be the required minimal distance $D(A, B)$.

The algorithm computes $D(i,j)$ by recursion. The value of $D(i,j)$ corresponds to the conversion of the initial subsequences $A_i = a_1a_2 \ldots a_i$, and $B_j = b_1b_2 \ldots b_j$ into a common sequence by L edit operations S_k, $k = 1,\ldots, L$, which can be considered to be applied in increasing order of position in the strings. Consider *undoing* the last of these edit operations. The resulting truncated sequence of edit operations, S_k, $k = 1,\ldots, L - 1$, is a sequence of edit operations for converting a substring of A_i and a substring of B_j into a common result. What is more, it must be an *optimal* sequence of edit operations for these substrings, for if some other sequence S'_k were a lower-cost sequence of operations for these substrings, then S'_k followed by S_L would be a lower-cost sequence of operations than S_k for converting A_i to B_j. Therefore, there should be a recursive method for calculating the $D(i,j)$.

Recognize the correspondence between individual edit operations and steps between adjacent squares in the matrix.

$(i-1, j-1) \to (i, j)$ corresponds to the substitution $a_i \to b_j$.
$(i-1, j) \to (i, j)$ corresponds to the deletion of a_i from A.
$(i, j-1) \to (i, j)$ corresponds to the insertion of b_j into A at position i.

Sequences of edit operations correspond to stepwise paths through the matrix

$$(i_0, j_0) = (0,0) \to (i_1, j_1) \to \ldots (n,m)$$

where $0 \le i_{k+1} - i_k \le 1$, (for $0 \le k \le n-1$), $0 \le j_{k+1} - j_k \le 1$ (for $0 \le k \le m-1$.) Considering the possible sequences of edit operations and the corresponding paths through the matrix, the predecessor of an optimal string of edit operations leading from (0,0) to $(i,\cdot j)$, where $i, j > 0$, must be an optimal sequence of edit operations leading to one of the cells $(i - 1, j)$, $(i - 1, j - 1)$, or $(i, j - 1)$; and, correspondingly, $D(i, j)$ must depend only on the values of $D(i - 1,j)$, $D(i--1, j-1)$ and $D(i, j-1)$, (together of course with the parameterization specified by the cost function d).

The algorithm is then as follows:

Compute the $(m + 1) \times (n + 1)$ matrix D by applying

1. the initialization conditions on the top row and left column:

$$D(i,0) = \sum_{k=0}^{i} d(a_k, \phi)$$

$$D(0,j) = \sum_{k=0}^{i} d(\phi, b_k)$$

These values impose the gap penalty on unmatched residues at the beginning of either sequence and then

2. the recurrence relations:

$D(i,j) = \min\{D(i\text{-}1,j)+d(a_i,\phi, D(i\text{-}1,j\text{-}1)+d(a_i,b_j), D(i, j\text{-}1)+d(f,b_j)\}$

for i=1,..., n; j=1,..., m. This means: consider all three possible steps to $D(i,j)$:

Operation	*Cumulative cost*
insert a gap in sequence A	$D(i - 1, j) + d(aj, \phi)$
substitute $a_i \leftrightarrow b_j$	$D(i - 1,j - 1) + d(a_i, b_j)$
insert a gap in sequence B	$D(i, j - 1) + d(\phi, b_j)$

From these, choose the minimal value. For each cell record not only the value $D(i, j)$ but a pointer back to (one or more of) the cell(s) $(i - 1, j)$, $(i - 1, j - 1)$ or $(i, j - 1)$ selected by the minimization operation. Note that more than one precedessor may give the same value.

When the calculations are complete, $D(n, m)$ is the optimal distance $D(A, B)$. An alignment corresponding to the sequence of edit operations recorded by the pointers can be recovered by tracing a path back through the matrix from (n, m) to (0,0). This alignment corresponding to the minimal distance $D(A,B) = D(n,m)$ may well not be unique.

•

Significance of Alignments

Suppose alignment reveals an intriguing similarity between two sequences. Is the similarity significant or could it have arisen by chance? For some simple phenomena—tossing a coin or rolling dice—it is possible to calculate exactly the expected distribution of results, and the likelihood of any particular result. For sequences it is not trivial to define the population from which the alignment is selected. For instance, to take random strings of nucleotides or amino acids as controls ignores the bias arising from non-random composition.

A practical approach to the problem is as follows: If the score of the alignment observed is no better than might be expected from a random permutation of the sequence, then it is likely to have arisen by chance. We may randomize one of the sequences, many times, realign each result to the second sequence (held fixed), and collect the distribution of resulting scores. For database searches, use the population of results returned from entire database as the population with which to measure the statistics.

Clearly, if the randomized sequences score as well as the original one, the alignment is unlikely to be significant. We can measure the mean and standard deviation of the scores of the alignments of randomized sequences, and ask whether the score of original sequence is unusually high. The Z-score reflects the extent to which the original result is an outlier from the population:

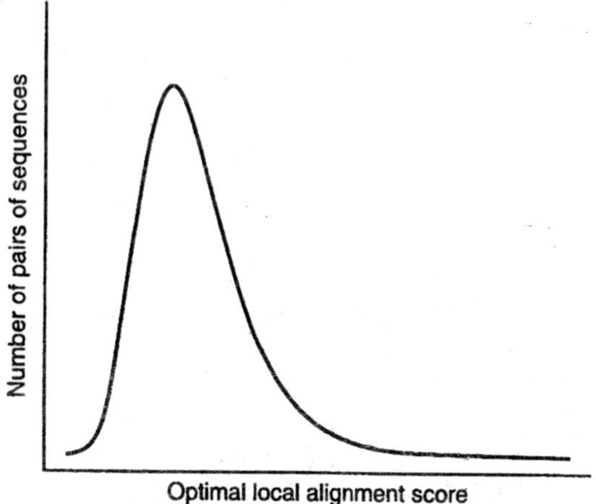

Fig. 2.2. Optimal local alignment scores for pairs of random amino acid sequences of the same length follow an extreme-value distribution.

$$Z - \text{score} = \frac{\text{score} - \text{mean}}{\text{standard deviation}}$$

A Z-score of 0 means that the observed similarity is no better than the average of random permutations of the sequence, and might well have arisen by chance. Other values used as measures of significance are P = the probability that the observed match could have happened by chance, and, for database searching, E = the number of matches as good as the observed one that would be expected to appear by chance in a database of the size probed.

Many 'rules of thumb' are expressed in terms of percent identical residues in the optimal alignment. If two proteins have over 45% identical residues in their optimal alignment, the proteins will have very similar structures, and are very likely to have a common or at least a similar function. If they have over 25% identical residues, they are likely to have a similar general folding pattern. On the other hand, observations of a lower degree of sequence similarity cannot rule out homology. R.F. Doolittle defined the region of 18—25% sequence identity as the 'twilight zone' in which the suggestion of homology is tantalizing but dangerous. Below the twilight zone is a region where pairwise sequence alignments tell very little. Lack of significant sequence similarity does not preclude similarity of structure.

Although the twilight zone is a treacherous region, we are not entirely helpless. In deciding whether there is a genuine relationship, the 'texture' of the alignment is important—are the similar residues isolated and scattered throughout the sequence; or are there 'icebergs'—local regions of high similarity (another term of Doolittle's), which may correspond to a shared active site? We may need to rely on other information about shared ligands or function. Of course, if the structures are known, we could examine them directly.

Some illustrative examples are:

1. Sperm whale myoglobin and lupin leghaemoglobin have 15% identical residues in optimal alignment. This is even below Doolittle's definition of the twilight zone. But we also know that both molecules have similar three-dimensional structures, both contain a haem group, and both bind oxygen. They are indeed distantly-related homologues.
2. The sequences of the N- and C-terminal halves of rhodanese have 11% identical residues in optimal alignment. If these

appeared in independent proteins, one could not conclude from the sequences alone that they were related. However, their appearance in the same protein suggests that they arose via gene duplication and divergence. The striking similarity of their structures confirms their relationship.

3. As a cautionary note, consider the proteinases chymotrypsin and subtilisin. They have 12% identical residues in optimal alignment. These enzymes have a common function, and a common catalytic triad. However, they have dissimilar folding patterns, and are not related. Their common function and mechanism is an example of convergent evolution. This case serves as a warning against special pleading for relationships between proteins with dissimilar sequences on the basis of similarities of function and mechanism!

MULTIPLE SEQUENCE ALIGNMENT

'One amino acid sequence plays coy; a pair of homologous sequences whisper; many aligned sequences shout out loud.' In nature, even a single sequence contains all the information necessary to dictate the fold of the protein. How does a multiple sequence alignment make the information more intelligible to us? Alignment tables expose patterns of amino acid conservation, from which distant relationships may be more reliably detected. Structure prediction tools also give more reliable results when based on multiple sequence alignments than on single sequences.

Visual examination of multiple sequence alignment tables is one of the most profitable activities that a molecular biologist can undertake away from the lab bench. Don't even THINK about not displaying them with different colours for amino acids of different physicochemical type. A reasonable colour scheme is:

Colour	*Residue type*	*Amino acids*
Yellow	Small nonpolar	Gly, Ala, Ser, Thr
Green	Hydrophobic	Cys, Val, lle, Leu, Pro, Phe, Tyr, Met, Trp
Magenta	Polar	Asn, Gln, His
Red	Negatively charged	Asp, Glu
Blue	Postively charged	Lys, Arg

To be informative a multiple alignment should contain a distribution of closely- and distantly-related sequences. If all the sequences are very closely related, the information they contain is largely redundant, and few inferences can be drawn. If all the sequences are very distantly related, it will be difficult to construct an accurate alignment (unless all the structures are available), and in such cases the quality of the results, and the inferences they might suggest, are questionable. Ideally, one has a complete range of similarities, including distantly-related examples linked through chains of close relationships.

STRUCTURAL INFERENCES FROM MULTIPLE SEQUENCE ALIGNMENTS

Thioredoxins are enzymes found in all cells. They participate in a broad range of biological processes, including cell proliferation, blood clotting, seed germination, insulin degradation, repair of oxidative damage, and enzyme regulation. The common mechanism of these activities is the reduction of protein disulphide bonds.

The structure of *E. coli* thioredoxin contains a central five-stranded β-sheet flanked on either side by α-helices; these helices and strands are indicated by the symbols α and β. Other thioredoxins are expected to share most but not all of the secondary structure of the *E. coli* enzyme. The plate also shows a summary of the alignment as a *sequence logo*, in which letters of different sizes indicate different proportions of amino acids.

Structural and functional features of thioredoxins that we might hope to identify from the multiple sequence alignment include

1. *The most highly conserved regions probably correspond to the active site.* The disulphide bridge between residues 32 and 35 in *E. coli* thioredoxin is part of a WCGPC[K or R] motif conserved in the family. Other regions conserved in the sequences, including the PT at residues 76—77 and the GA at residues 92—93, are involved in substrate binding.
2. *Regions rich in insertions and deletions probably correspond to surface loops. A position containing a conserved Gly or Pro probably corresponds to a turn.* Turns correlated with insertions and deletions occur at positions 9, 20,60 and 95. The conserved glycine at position 92 in *E. coli* thioredoxin is indeed part of a turn. It is in an unusual mainchain conformation, one that is easily accessible only to glycine. The conserved proline at position 76 in *E. coli*

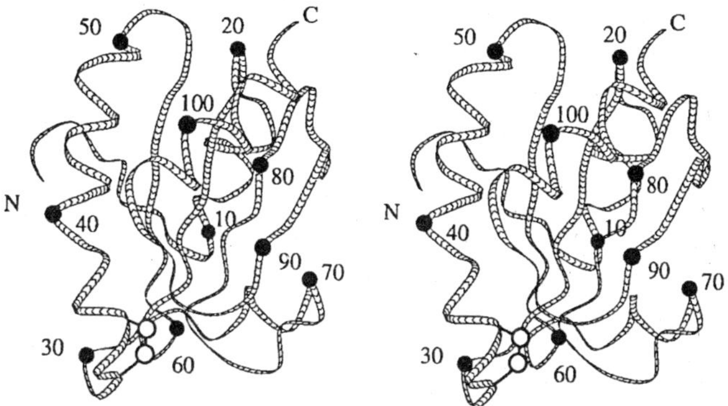

Fig. 2.3. The structure of E. coli thioredoxin.

thioredoxin is also associated with a turn. It is in another unusual mainchain conformation, this one easily accessible only to proline.

3. *A conserved pattern of hydrophobicity with spacing 2 (that is, every other residue)—with the intervening residues more variable and including hydrophilic residues—suggests a β-strand on the surface.* This pattern is observable in the β-strand between residues 50 and 60.
4. *A conserved pattern of hydrophobicity with spacing ~4 suggests a helix.* This pattern is observable in the region of helix between residues 40 and 49.

Thioredoxins are members of a superfamily including many more distantly-related homologues. These include glutaredoxin (hydrogen donor for ribonucleotide reduction in DNA synthesis), protein disulphide isomerase (which catalyses exchange of mismatched disulphide bridges in protein folding), phosducin (a regulator of G-protein signalling pathways), and glutathione S-transferases (chemical defense proteins). Implicit in the multiple sequence alignment table of the thioredoxins themselves are patterns that should be applicable to identifying these more distant relatives.

Applications of Multiple Sequence Alignments to Database Searching

Searching in databases for homologues of known proteins is a central theme of bioinformatics. Indeed it brooked no delay; with the application of PSI-BLAST. We reconsider database searching here, with the goal of trying to understand how we can best use available information to build effective procedures. The goals are high *sensitivity*—picking up even very distant relationships—and high *selectivity*—minimizing the number of sequences reported that are not true homologues.

We recognize a familiar face by reacting to its integral appearance rather than to individual features. Similarly, multiple sequence alignments contain subtle patterns that characterize families of proteins.

During the last decade, great progress has been made in devising methods for applying multiple sequence alignments of known proteins to identify related sequences in database searches. The results are central to contemporary applications of bioinformatics, including the interpretion of genomes. Three important methods are: Profiles, PSI-BLAST and Hidden Markov Models (HMMs).

Profiles

Profiles express the patterns inherent in a multiple sequence alignment of a set of homologous sequences. They have several applications:

1. They permit greater accuracy in alignments of distantly-related sequences.
2. Sets of residues that are highly conserved are likely to be part of the active site, and give clues to function.
3. The conservation patterns facilitate identification of other homologous sequences.
4. Patterns from the sequences are useful in classifying subfamilies within a set of homologues.
5. Sets of residues that show little conservation, and are subject to insertion and deletion, are likely to be in surface loops. This information has been applied to vaccine design, because such regions are likely to elicit antibodies that will cross-react well with the native structure.
6. Most structure-prediction methods are more reliable if based on a multiple sequence alignment than on a single sequence. Homology modelling, for instance, depends crucially on correct sequence alignments.

To use profile patterns to identify homologues, the basic idea is to match the query sequences from the database against the sequences in the alignment table, giving higher weight to positions that are conserved than to those that are variable. If a region is absolutely conserved, such as the WGCPC motif in thioredoxins, the procedure should all but insist on finding it. But the risk of being too compulsive is to miss interesting distant relatives; some leeway should be allowed.

What is needed is a quantitative measure of conservation. For each position in the table of aligned sequences, take an inventory of the distribution of amino acids. For instance, for positions 25–30 of the thioredoxin alignment:

Residue number	*Number of each amino acid*																			
	A	C	D	E	F	G	H	I	K	L	M	N	P	Q	R	S	T	V	W	Y
25	1									2								13		
26			16																	
27					16															
28																7	1		5	3
29	16																			
30			1	4									2			1	7	1		

Given a query sequence representing a potential thioredoxin homologue, we want to evaluate its similarity to the query sequence, in such a way that agreement with the known sequences at the absolutely conserved positions—for instance 26,27 and 29— contributes a very high score, and disagreement at these positions contributes a very low score. For moderately conserved positions, such as 28, we want a modest positive contribution to the score if the query sequence has an S or a W at this position, and a smaller contribution if it has T or Y. The general idea is to score each residue from the query sequence based on the amino acid distribution at that position in the multiple sequence alignment table.

A tempting but overly simple approach would be to use the inventories as scores directly. For example, if the region in a query sequence that corresponds to positions 25–30 in thioredoxin contains the sequence VDFSAE, this fragment would score 13+16+16+6+16+4=71. This is almost the greatest value possible. The alternative query sequence ACGVAP would score 1+0+0+5+16+2=24, a much lower value. Of course, for each query sequence we have to test all possible alignments with the multiple alignment table, and take the largest total score. The highest-scoring sequences best fit the patterns implicit in the table.

This simple approach would work if our table contained a large and unbiased sample of thioredoxin sequences. But only in this case would the simple inventory give a correct picture of the *potential* distribution of residues at each position. If our sample were small, the pattern derived would be unlikely to reflect the complete repertoire. Or, if the sample contained a large subset of similar sequences, these would be overrepresented in the inventories. For instance we can see in Plate VII that vertebrate thioredoxins form a very closely related set. If we included twenty more vertebrate thioredoxins in the alignment, the profile would recognize only vertebrate thioredoxins effectively.

Substitution matrices suggest how to make the inventory 'fuzzy' and thereby more general.

The observed amino acid distribution at any residue position is a 20-membered array: $(a_1, a_2, a_3, ..., a_{20})$, where a_i is the number of amino acids of type i observed at that position. (For position 25 of the thioredoxins, $a_1 = 1$ because 1 alanine is observed, and $a_{18} = 13$, representing the valines.) Then, in the simplest scheme, the score of an alanine at position 25 is just 1; the score of a Val is 13; in general, the score of an amino acid of type i is a_i. In this scheme the rows of inventory itself provide the arrays a needed for scoring each position.

A better scoring scheme would evaluate any amino acid according to its chance of being subtituted for one of the observed amino acids. If $D(i, j)$ is the amino acid substitution matrix — PAM250 or BLOSUM62, for example — then amino acid i could score $a_1D(i, 1) + a_2D(i, 2)... a_{20}D(i, 20)$. This scheme distributes the score among observed amino acids, weighted according to the substitution probability. An amino acid in the query sequence could score high either if it appears frequently in the inventory at this position, or if it has a high probability of arising by mutation from residue types that are common at this position. This approach is more effective in detection of distant relatives from a limited set of known sequences. In this case, the scoring vector for amino acids is the product of the substitution matrix and the rows of the inventory array. An even better approach is to use as the amino acid distribution a combination of the observed inventory and a general background level of amino acid composition.

The result is a set of probability scores for each amino acid (or gap) at each position of the alignment, called a *position-specific scoring matrix*. Given a query sequence, and the position-specific scoring matrices derived from a profile, the calculations required to find the optimal score over all alignments of the query sequence with the profile are extensions of the dynamic programming methods of aligning two sequences.

A weakness of simple profiles is that the multiple sequence alignment must be provided in advance, and is taken as fixed. PSI-BLAST and Hidden Markov Models gain power by integrating the alignment step with the collection of statistics.

PSI-BLAST

PSI-BLAST is a program that searches a databank for sequences similar to a query sequence. It is a development of the earlier program BLAST = Basic Local Sequence Alignment Tool. The BLAST program and its variants check each entry in the databank *independently* against a query sequence. PSI-BLAST begins with such a one-at-a-time search. It then derives pattern information from a multiple sequence alignment of the initial hits, and reprobes the database using the pattern. Then it repeats the process, fine-tuning the pattern in successive cycles.

The problem that BLAST was originally designed to solve is that full-blown dynamic programming methods are rather slow for complete searches in a large databank. Often the databank contains close matches to the query sequence. Less sensitive but faster programs are quite capable of identifying the close matches, and if that is what is required, fine. For example, if one wants to search for homologues of a mouse protein in the human genome, the similarity is likely to be high and an approximate method likely to find it. But if one wants to search for homologues of a human protein in *C. elegans* or yeast, the relationship may be more tenuous and more sophisticated, slower methods may be required. (It may come as a surprise, but computer time requirements are still a consideration. For although computing is becoming less expensive, the sizes of the databanks and the number of searches desired, on a worldwide basis, are growing. The net effect is that the pressure on computing resources is increasing.)

The method used by BLAST goes back, in a sense, to the dotplot approach, checking for well-matching local regions. For each entry in the database, it checks for short contiguous regions that match a short contiguous region in the query sequence, using a substitution scoring matrix but allowing no gaps. An approach in which candidate regions of *fixed length* are identified initially can be made very fast by the use of lookup tables.

Once BLAST identifies a well-fitting region, it tries to extend it. In some versions gaps are allowed. Even a very simple algorithm could pick up all matching regions of five contiguous residues and then combine and extend them.

PSI-BLAST, using iterated pattern search, is much more powerful than simple pairwise BLAST in picking up distant relationships. PSI-BLAST correctly identifies three times as many homologues as BLAST in the region below 30% sequence identity. It is therefore a very useful method for analysing whole genomes. PSI-BLAST was able to match protein domains of known structure to 39% of the genes in *M. genitalium,* 24% of the genes in yeast, and 21% of the genes in *C. elegans*. The only methods based entirely on sequence analysis that do better than PSIBLAST are Hidden Markov Models. These are described in the next section. To achieve significantly better performance it is necessary to make explicit use of structural information.

Hidden Markov Models (HMMs)

A Hidden Markov Model is a computational structure for describing the subtle patterns that define families of homologous sequences. HMMs are powerful tools for detecting distant relatives, and for prediction of protein folding patterns. They are the only methods based entirely on sequences—that is, without explicitly using structural information—competitive with PSI-BLAST for identifying distant homologues. They also perform well at fold recognition, as assessed in CASP programs.

Within an HMM is a multiple sequence alignment. However, HMMs are usually presented as *procedures for generating sequences*. A conventional multiple sequence alignment table could also be used to generate sequences, by selecting amino acids at successive positions, each amino acid chosen from a position-specific probability distribution derived from the profile. But HMMs are more general than profiles.

1. They include the possibility of introducing gaps into the generated sequence, with position-dependent gap penalties.
2. Application of profiles requires that the multiple sequence alignment be specified up front; the pattern statistics are then

derived from the alignment. HMMs carry out the alignment and the assignment of probabilities together.

The internal structure of an HMM shows the mechanism for generating sequences. Begin at Start, and follow some chain of arrows until arriving at End. Each arrow takes you to a state of the system. At each state (1) you take some action—emit a residue, perhaps—and (2) choose an arrow to take you to the next state. The action and the choice of successor state are governed by sets of probabilities. Associated with each state that emits a residue are: one probability distribution for the twenty amino acids, and a second probability distribution for the choice of successor state. Both of these probability distributions are calibrated to encode information about a particular sequence family. In this way, the same general mathematical framework can be specialized to many different sequence families.

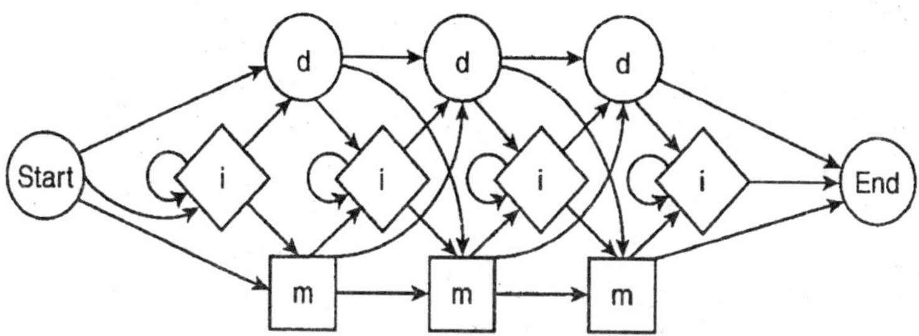

Fig. 2.4. The structure of a Hidden Markov Model (HMM).

The dynamics of the system are such that only the current state influences the choice of its successor—the system has no 'memory' of its history. This is characteristic of the processes studied by the nineteenth-century Russian mathematician A.A. Markov. Distinguish the succession of states from the succession of amino acids emitted to form the output sequence. Several paths through the system can generate the same sequence. Only the succession of characters emitted is visible; the state sequence that generated the characters remains internal to the system, that is, hidden. By the probability distributions associated with the individual states the system captures—or models—the patterns inherent in a family of sequences. Hence the name, Hidden Markov Model.

Software for applying HMMs to biological sequence analysis can achieve

1. *Training*. Given a set of unaligned homologous sequences, it can align them and adjust the transition and residue-output probabilities to define an HMM capturing the patterns inherent in the sequences submitted.
2. *Detection of distant homologues*. Given an HMM and a test sequence, calculate the probability that the HMM would generate the test sequence. If an HMM trained on a known family of sequences would generate the test sequence with relatively high probability, the test sequence is likely to belong to the family.
3. *Align additional seauences*. The probability of any sequence of states can be computed from the individual state-to-state transition probabilities. Finding the most likely sequence of states that the HMM would use to generate one or more test sequences reveals their optimal alignment to the family.

Phylogeny

We have now seen several examples of evolution, in proteins and in genomes. These represent the extension to the molecular level of concerns that have occupied biologists since Darwin and even before. The basic principle is that *the origin of similarity is common ancestry.* Although there are many exceptions, arising from convergent evolution, the importance of this principle both for rationalizing contemporary observations and giving a window into the hsitory of life cannot be overestimated.

The field of phylogeny has the goals of working out the relationships among species, populations, individuals, or genes. Relationship is taken in the literal sense of kinship or genealogy, that is, assignment of a scheme of descendants of a common ancestor. The results are usually presented in the form of an evolutionary tree. The taxonomy of the ratites—large flightless birds—is a typical example. The ancestor of the ratites is believed to be a bird that could fly, probably related to the extant tinamous.

Such a tree, showing all descendants of a single original ancestral species, is said to be *rooted*. (The root of the tree typically appears at the top or the side; botanists will have to get used to this.) Alternatively, we may be able to specify relationships but not order them according to a history. The relationships among the finches of the Galapagos Islands, studied by Darwin, plus a related species from the nearby Cocos Island, are shown in an

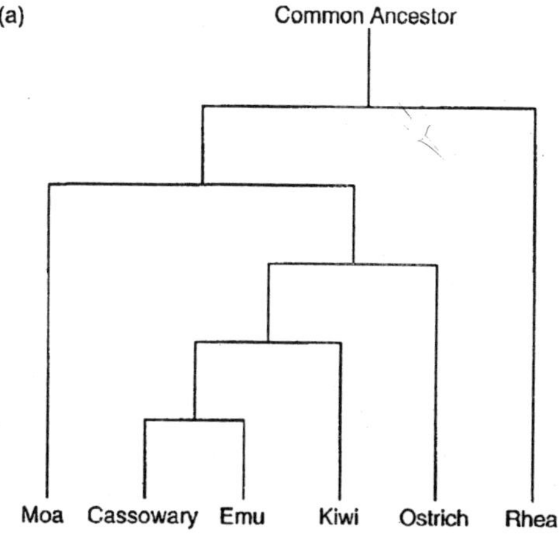

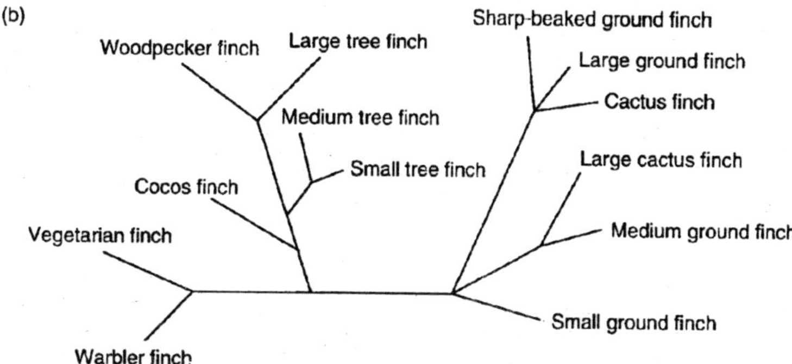

Fig. 2.5. (a) Phylogenetic tree of ratites (large flightless birds) based on mitochondrial DNA sequence. (b) Unrooted tree of relationships among finches from the Galapagos and Cocos Island.

unrooted tree. Addition of data from a species on the South American mainland ancestral to the island finches would allow us to *root the tree.*

Statement of a tree of relationships may reveal only the connectivity or topology of the tree, in which case the lengths of the branches contain no information. A more ambitious goal is to show the distances between taxa quantitatively; for instance, to label the branches with the time since divergence from a common ancestor.

Given a set of data that characterize different groups of organisms—for example, DNA or protein sequences, or protein structures, or shapes of teeth from different species of animals—

how can we derive information about the relationships among the organisms in which they were observed? It is rare for species relationships and ancestry to be directly observable. Evolutionary trees determined from genetic data are often based on inferences from the patterns of similarity, which are all that is observable among species living now. We generally assume that the more similar the characters the more closely related the species, although this is a dangerous assumption. Nevertheless, from the relationships among the characters we wish to infer patterns of ancestry: the *topology* of the phylogenetic realtionships (informally, the 'family tree.')

To what extent do the topologies of the relationships depend on the choice of character? In particular, are there *systematic* discrepancies between the implications of molecular and palaeontological analysis?

Molecular approaches to phylogeny developed against a background of traditional taxonomy, based on a variety of morphological characters, embryology, and, for fossils, information about the geological context (stratigraphy). The classical methods have some advantages. Traditional taxonomists have much less restricted access to extinct organisms, via the fossil record. They can *date* appearances and extinctions of species by geological methods. Molecular biologists, in contrast, have very limited access to extinct species. Some sub-fossil remains of species which became extinct as recently as the last century or two have legible DNA, including specimens of the quagga (a relative of the zebra) and the thylacine (Tasmanian 'wolf', a marsupial), and some New Zealand birds (including moas). We have already seen an example of a sequence from the mammoth. Some DNA sequences from Neanderthal man have been recovered from an individual who died approximately 30,000 years ago. But *Jurassic Park* remains fiction!

A crucial event in the acceptance of molecular methods occurred in 1967 when V.M. Sarich and A.C. Wilson dated the time of divergence of humans from chimpanzees at 5 million years ago, based on immunological data. At that time palaeontologists dated this split at 15 million years ago, and were reluctant to accept the molecular approach. Reinterpretation of the fossil record led to acceptance of a more recent split, and broke the barrier to general acceptance of molecular methods.

Indeed, many molecular properties have been used for phylogenetic studies, some surprisingly long ago. Serological cross-reactivity was used from the beginning of the last century until superseded by direct use of sequences. In one of the most premature scientific studies I know of, E.T. Reichert and A.P. Brown published, almost a century ago (in 1909), a phylogenetic analysis of fishes based on haemoglobin crystals. Their work was based on Stenö's law (1669), that although different crystals of the same substance have different dimensions—some are big, some small —they have the same interfacial angles, reflecting the similarity in microscopic arrangement and packing of the atomic or molecular units within the crystals. Reichert and Brown showed that the interfacial angles of crystals of haemoglobins isolated from different species showed patterns of similarity and divergence parallel to the species' taxonomic relationships.

Reichert and Brown's results are replete with significant implications. They show that proteins have definite, fixed shapes, an idea by no means recognized at the time. They imply that as species progressively diverge, the structures of their haemoglobins progressively diverge also. In 1909, no one had a clue about nucleic acid or protein sequences. In principle, therefore, the recognition of evolution of protein structures preceded, by several decades, the idea of evolution of sequences.

Today, DNA sequences provide the best measures of similarities among species for phylogenetic analysis. The data are digital. It is even possible to distinguish selective from non-selective genetic change, using the third position in codons, or untranslated regions such as pseudogenes, or the ratio of synonymous to non-synonymous codon substitutions. Many genes are available for comparison. This is fortunate, because, given a set of species to be studied, it is necessary to find genes that vary at an appropriate rate. Genes that remain almost constant among the species of interest provide no discrimination of degrees of similarity. Genes that vary too much cannot be aligned. There is an analogous situation in radioactive dating requiring choice of an isotope with a half-life of the same general magnitude as the time interval to be determined.

Fortunately, genes vary widely in their rates of change. The mammalian mitochondrial genome, a circular double-stranded DNA molecule approximately 16,000 bp long, provides a useful fast-changing set of sequences for study of evolution among closely-related species. In contrast, ribosomal RNA sequences were used by C. Woese to identify the three major divisions: Archaea, Bacteria and Eukarya.

Conversely, different rates of change of sequences of different genes can lead to different and even contradictory results in phylogenetic studies. This is especially true if what we want is not just the topology of the relationships but the branch lengths. In addition, horizontal gene transfer, and convergent evolution, are competing phenomena—that is, competing with descent—that interfere with the deduction of phylogenetic relationships.

Phylogenetic Trees

We describe phylogenetic relationships as trees. In computer science, a tree is a particular kind of graph. A graph is a structure containing nodes (abstract points) connected by edges (represented as lines between the points). A *path* from one node to another is a consecutive set of edges beginning at one point and ending at the other, like our trip from Malmö to Tromsø. A *connected graph* is a graph containing at least one path between any two nodes. From these we can define a *tree*: a connected graph in which there is *exactly* one path between every two points. A particular node may be selected as a *root*; but this is not necessary —abstract trees may be rooted or unrooted. Unrooted trees show the topology of relationship but not the pattern of descent. A rooted tree in which every node has two descendants is called a *binary tree*.

Another special kind of graph is a *directed graph* in which each edge is a one-way street. Rooted phylogenetic trees are, implicitly, directed graphs, the ancestor-descendent relationship implying the direction of each edge.

It may be possible to assign numbers to the edges of a graph to signify, in some sense, a 'distance' between the nodes connected by the edges. The graph may then be drawn to scale, with the sizes of the edges proportional to the assigned lengths. The length of a path through the graph is the sum of the edge lengths.

In phylogenetic trees, edge lengths signify either some measure of the dissimilarity between two species, or the length of time since their separation. The assumption that differences between properties of living species reflects their divergence times will be true only if the rates of divergence are the same in all branches of the tree. Many exceptions are known; for instance, among mammals rodents show relatively fast evolutionary rates for many proteins.

Broadly, there are two approaches to deriving phylogenetic trees. One approach makes no reference to any historical model of the relationships. Proceed by measuring a set of distances between species, and generate the tree by a hierarchical clustering procedure. This is called the *phenetic* approach. The alternative, the *cladistic* approach, is to consider possible pathways of evolution, infer the features of the ancestor at each node, and choose an optimal tree according to some model of evolutionary change. Phenetics is based on similarity; cladistics is based on genealogy.

Clustering Methods

Phenetic or clustering approaches to determination of phylogenetic relationships are explicitly non-historical. Indeed, hierarchical clustering is perfectly capable of producing a tree even in the absence of evolutionary relationships. A departmental store has goods clustered into sections according to the type of product—for instance, clothing or furniture—and subclustered into more closely related subdepartments, such as men's and women's shoes. Men's and women's shoes have a common ancestor, but there is no implication that shoes and furniture do.

A simple clustering procedure works as follows: given a set of species, determine for all pairs a measure of the similarity or difference between them. This could depend on a physical body trait such as the difference between the average adult height of members of two species. Or one could use the number of different bases in aligIlments of mitochondrial DNA. To create a tree from the set of dissimilarities, first choose the two most closely related species and insert a node to represent their common ancestor. Then replace the two selected species by a set containing both, and replace the distances from the pair to the others by the average of the distances of the two selected species to the others. Now we have a set of pairwise dissimilarities, not between individual species, but between sets of species.

This process of tree building is called the UPGMA method (Unweighted Pair Group Method with Arithmetic mean). A modification of the UPGMA method by N. Saitou and M. Nei, called Neighbour Joining, is designed to correct for unequal rates of evolution in different branches of the tree.

Cladistic Methods

Cladistic methods deal explicitly with the patterns of ancestry implied by the possible trees relating a set of taxa. Their aim is to select the correct tree by utilizing an explicit model of the evolutionary process. The most popular cladistic methods in molecular phylogeny are the *maximum parsimony* and *maximum likelihood* approaches. They are specialized to sequence data, starting from a multiple sequence alignment. Neither maximum parsimony nor maximum likelihood could be applied to anatomic characters such as average adult height.

The *maximum parsimony* method of W. Fitch defines an optimal tree as the one that postulates the fewest mutations. For instance, given species characterized by homologous sequences ATCG, ATGG, TCCA, and TTCA, the tree:

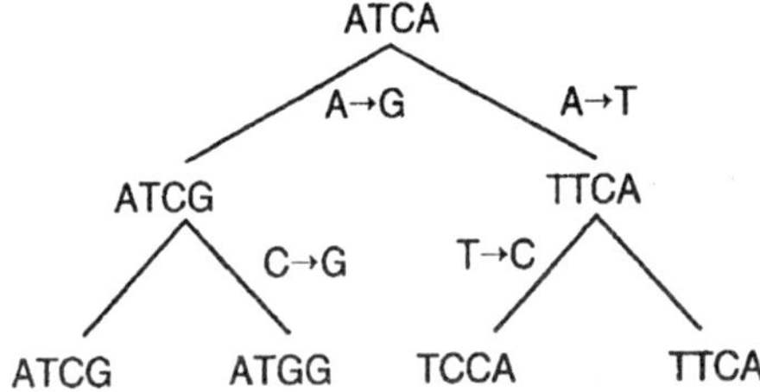

postulates 4 mutations. An alternative tree:

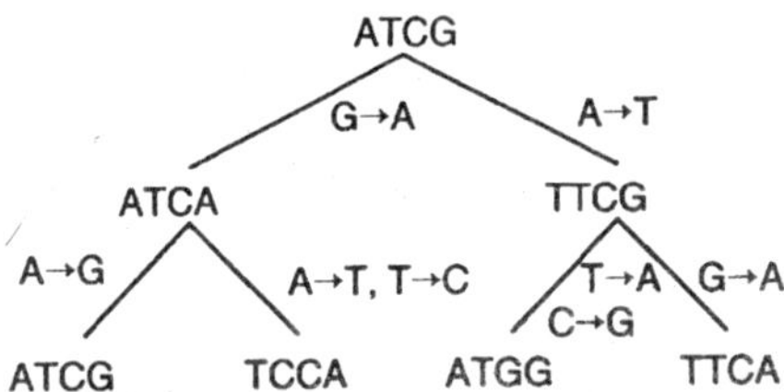

postulates 7 mutations. Note that the second tree implies that the G→A mutation in the fourth position occurred twice independently. The former tree is optimal according to the maximum parsimony method, because no other tree involves fewer mutations. In many cases, several trees may postulate the same number of mutations, fewer than any other tree. For such cases the maximum parsimony approach does not give a unique answer.

The *maximum likelihood* method assigns quantitative probabilities to mutational events, rather than merely counting them. Like maximum parsimony, maximum likelihood reconstructs ancestors at all nodes of each tree considered; but it also assigns branch lengths based on the probabilities of the mutational events postulated. For each possible tree topology, the assumed substitution rates are varied to find the parameters that give the highest likelihood of producing the observed sequences. The optimal tree is the one with the highest likelihood of generating the observed data.

Both maximum parsimony and maximum likelihood methods are superior to clustering techniques. This has been demonstrated with cases where independent evidence—for instance, from classical palaeontology—provides a correct answer, and also with simulated data—computed generation of evolving sequences.

Problem of Varying Rates of Evolution

Suppose that the four species A, B, C, and D have the phylogenetic tree:

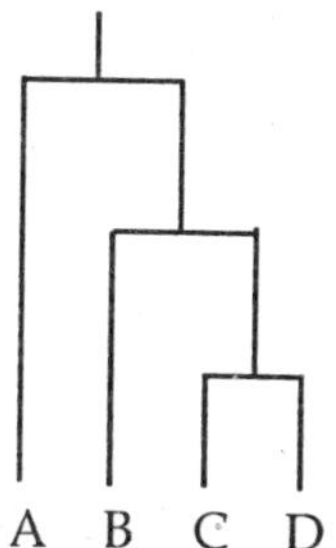

This tree is consistent with the dissimilarity matrix:

	A	B	C	D
A	0	3	3	3
B		0	2	2
C			0	1
D				0

Suppose however that species D is changing very fast, although the phylogeny is unaltered. The dissimilarity matrix might then be observed to be:

	A	B	C	D
A	0	3	3	20
B		0	2	20
C			0	20
D				0

from which we would derive the incorrect phylogenetic tree:

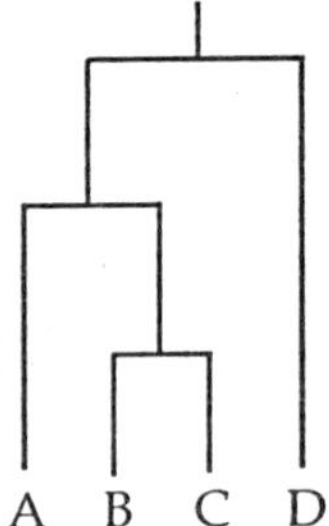

All the methods discussed here are subject to errors of this kind if the rates of evolutionary change vary along different branches of the tree. To test for varying rates, compare the species under consideration with an *outgroup*—a species more distantly related to all the species in question than any pair of them is to each other. For instance, if we are studying species of primates, a nonprimate mammal such as the cow would be a suitable outgroup. If the rates of evolution among the primate species were constant, we should expect to observe approximately equal dissimilarity measures between all primate species and the cow. If this is not observed, the suggestion is that evolutionary rates have varied among the primates, and the character being used may well not provide the correct phylogenetic tree.

Computational Considerations

Cladistic methods—maximum parsimony and maximum likelihood—are more accurate than simpler clustering methods such as UPGMA, but require large amounts of computer time if the number of species is appreciable. The total number of possible trees, which cladistic methods are committed to considering if they could, increases very rapidly with the number of species. As a result, in many cases of interest these methods can give only approximate answers, even with respect to their intrinsic assumptions.

Because calculated phylogenies are often approximations, it is important to try to test them. Methods include:

1. Comparison of phylogenies obtained from different characters describing the same set of taxa—are they consistent? If trees produced from different characters share a subtree, perhaps that portion of the phylogeny has been determined reliably and other portions have not.

2. Analysis of subsets of taxa should give the same answer—with respect to the subset—as appears within the full tree.
3. Formal statistical tests, involving re-running the calculation on subsets of the original data, are known as *jackknifing* and *bootstrapping*:

   Jackknifing is calculation with data sets sampled randomly from the original data. For phylogeny calculations from multiple sequence alignments, select different subsets of the positions in the alignment, and rerun the calculation. Finding that each subset gives the same phylogenetic tree lends it credibility. If each subset gives a different tree, none of them is trustworthy.

   Bootstrapping is similar to jackknifing except that the positions chosen at random may include multiple copies of the same position, to form data sets of the same size as the original, to preserve statisti cal properties of the sampling.
4. If there are very long edges, consider seriously the possibility of unequal variation in evolutionary rate that may have perturbed the calculation. Introduce outgroup taxa to check.

3

DATABASE CONTENTS

DNA SEQUENCING

Principles of DNA Sequencing

The order of nucleotides in DNA is determined by *chain termination sequencing* (also called *dideoxy sequencing* or the *Sanger method* after its inventor). The basic sequencing reaction, which consists of a *single-stranded DNA template*, a primer to initiate the nascent chain, four *deoxyribonucleoside triphosphates* (dATP, dCTP, dGTP and dTTP) and the enzyme *DNA polymerase*, which inserts the complementary nucleotides in the nascent DNA strand using the template as a guide. Four DNA polymerase reactions are set up, each containing a small amount of one of four *dideoxyribonucleoside triphosphates* (ddATP, ddCTP, ddGTP or ddTTP). These act as chain-terminating competitive inhibitors of the reaction, but are present in limiting amounts so there is only a small chance the growing chain will terminate at any given position. Therefore, each of the four reaction mixtures generates a *nested set* of DNA fragments, each terminating at a specific base. Nowadays, most DNA sequencing reactions are *automated*. Each reaction mixture is labeled with a different fluorescent tag (on either the primer or on one of the nucleotide substrates), which allows the terminal base of each fragment to be identified by a scanner.

All four reaction mixtures are then pooled and the DNA fragments separated by polyacrylamide gel electrophoresis (PAGE). Smaller DNA fragments travel through the gel faster than larger ones so the nested DNA fragments are separated according

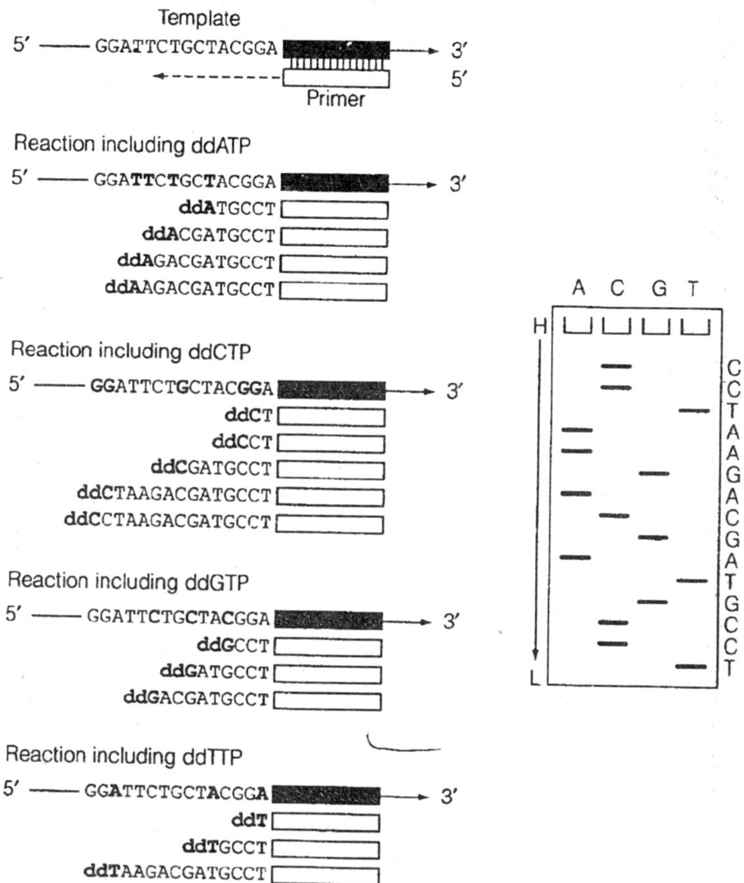

Fig. 3.1. Principle of DNA sequencing. (a) Four sequencing reactions are set up, each containing a limiting amount of one of the four dideoxynucleotides. (b) A polyacrylamide gel is shown with each reaction running in a separate lane for clarity.

to size. The resolution of PAGE-allows polynucleotides differing in length by only one residue to be separated. Near the bottom of the gel, the scanner 'reads' the fluorescent tag as each. DNA fragment moves past, and this is converted into *trace data,* displayed as a graph comprising coloured peaks corresponding to each base.

Types of DNA Sequence

DNA sequences are stored in databases such as GenBank and are generally of three types: genomic DNA, cDNA and

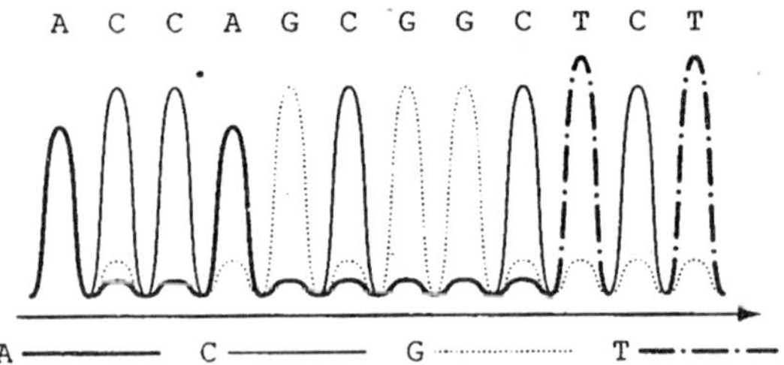

Fig. 3.2. A sample of a high-quality sequence trace, where all peaks are easily called. Peaks are typically printed in different colours to aid visual interpretation.

recombinant DNA. *Genomic DNA* is taken direct from the genome and contains genes in their natural state, which in eukaryotes includes introns, regulatory elements and large amounts of surrounding intergenic DNA. In contrast, *copy DNA* (*cDNA*) is prepared by reverse transcribing messenger RNA (mRNA). It is often useful to focus on cDNA because this provides direct access to genes, which may be difficult to find in genomic DNA.

In the human genome for example, genes represent only 3% of the sequence. *Recombinant DNA* includes the sequences of vectors such as plasmids, modified viruses and other genetic elements that are used in the laboratory. It is essential for these sequences to be stored in databases because linked vector sequences must be removed from new genomic and cDNA sequences prior to alignment to avoid spurious matches.

Genome Sequencing Strategies

Using current technology, only about 800 nucleotides of DNA sequence can be read in a single reaction. If a larger DNA molecule, such as an entire genome, is to be sequenced, it must first be broken up into smaller pieces, each of which has to be sequenced individually. There are two strategies for assembling genome sequences from short reads. In the *shotgun sequencing* strategy, the large DNA molecule is randomly broken up and many sequencing reactions are performed to cover the entire molecule. The entire sequence is reassembled by using a computer to search for overlaps. This method generates a large amount of sequence data very quickly, but difficulty may be experienced in closing the final gaps, a process termed *finishing*.

In the alternative *clone contig* approach, DNA fragments are subcloned in a rational manner and systematically sequenced until the entire sequence is completed. Using this method, sequence data accumulates more slowly at the beginning of a project, but there are fewer sequence gaps at the end so finishing is easier.

Sequence Quality Control

Genome sequencing is subject to strict *quality control* measures. For each clone, the sequencing reaction is performed several times on each strand. This primary trace data is then subject to *pre-assembly processing,* which includes base calling and quality assessment, the removal of vector sequences and the masking of repeats. Finally, the sequence reads are assembled into *contigs* (contiguous sequence reads). *Base calling* means deciding to which base a particular trace peak corresponds. Ideally, each peak is clear and unambiguous, but the quality of trace data can vary considerably.

Several computer programs are available that automatically make base calls from sequence traces, and assess the quality (i.e. confidence) of each call. One of the most widely used utilities is *Phred,* which uses dynamic programming to match observed and predicted peak locations and evaluate the quality of the trace data for each base at each position. The next processing step is to remove vector sequences. Sequence reads usually start with part of the sequencing vector and it is important to remove this sequence because it can generate false overlaps, Programs such as *vector_clip* and *CrossMatch* have been developed which align new sequence reads to a database of known vector sequences, and then mask off the vector sequence from further analysis.

An essential step in genome sequence assembly is the identification of repetitive DNA elements. Genomic DNA is rich in repetitive elements, which can be arranged in tandem or dispersed throughout the genome. Such repeats are extremely troublesome when it comes to reassemble individual reads into contigs because they generate false overlaps. The program *RepeatMasker* is widely used to identify and tag such elements. Finally, the assembly process itself, which is carried out using programs such as *Phrap*. This is a computationally difficult task owing to the large number of possible overlaps. For example, if there are 100 reads, there are nearly 40 000 different arrangements.

The Phrap assembly method is iterative and based on calculating the log likelihood ratio for each possible match taking into account both the sequence and the trace quality data. Despite the high level of automation, human curation to resolve disputed base calls and other discrepancies is required. Editing and contig assembly can be carried out using programs such as *Staden Gap4*, which allow trace data to be evaluated and the results of multiple sequence reads to be displayed.

Single-pass Sequencing

As discussed above, multiple sequencing reactions on both strands of a cloned DNA molecule are usually required to insure accuracy. However, where large amounts of sequence data are generated (e.g. in genome projects) *single-pass sequencing* is quicker, cheaper and provides more data at the expense of some accuracy. GenBank has subsidiary databases for such sequences. For example, *dbEST* is a database for *expressed sequence tags* (*ESTs*), which are generated by rapid, single-pass sequencing of random clones from cDNA libraries. Although short (2-300 bp) and inaccurate, very large numbers of sequences have been collected rapidly and inexpensively, and can be used to identify genes in genomic DNA and to prepare large clone sets for DNA microarrays.

The genomic equivalent of dbEST is *dbGSS*, a database for *genome survey sequences* (*GSSs*), which are random, single-pass sequences of genomic DNA. Large amounts of preliminary genomic sequence data (*unfinished sequence*) are stored in the *HTG* (*high-throughput genomic*) division of GenBank. Note that the vast majority of ESTs are eukaryotic sequences. This is because cDNA synthesis is generally carried out using a primer that anneals to the *polyadenylate tail* of the eukaryotic mRNA. Most bacterial mRNAs lack a polyadenylate tail and cannot be copied using the same strategy. Furthermore, many bacterial transcripts are never synthesized as full-length molecules due to ,their operon organization and the coupling of transcription and translation. However, bacteria also tend to have compact genomes and genes that lack introns, therefore genomic DNA fragments are already very similar to cDNAs.

RNA Sequencing

RNA sequencing is less straightforward than DNA sequencing since there are large numbers of *minor nucleotides* (chemically modified nucleotides), for example in transfer RNA (tRNA) and ribosomal RNA (rRNA). An RNA sequence lacking data on minor nucleotides can be deduced from either the DNA sequence of the corresponding gene or from a cDNA sequence (the sequence of the reverse transcript). In many cases, cDNA sequences are more informative than gene sequences because the RNA molecule may be extensively processed during synthesis. For example, *introns* may be spliced out of a primary transcript to generate a mature mRNA, and extra nucleotides may be inserted by *RNA editing*. Direct RNA sequencing involves the chemical characterization of modified nucleotides. One type of modification, 2′-*O*-methylation, is easy to pick up as the phosphodiester bond on the 3'-side of such a nucleotide is not alkali-labile. Other types of modification can be identified by mass spectrometry.

Protein Sequencing

Protein sequences can be deduced from DNA sequences, but as with RNA sequencing this does not provide information on modified residues or other types of post-translational protein modification (such as cleavage or the formation of disulfide bonds). In the past, direct protein sequencing was carried out using a process called *Edman degradation*, in which the terminal residue of a protein was labeled, removed and then identified using a series of chemical tests. Current methods for protein sequencing rely on *mass spectrometry* (*MS*), a technique in which the *mass/charge ratio (m/e* or *m/z)* of ions in a vacuum is accurately determined, allowing molecular masses to be calculated at a resolution in the order of 10^{-5}. Only extremely small quantities of material are required for analysis and the resolution of the technique is such that, for small molecules, the accurate *m/e* value directly reveals the chemical formula. Hence, the technique can be used to characterize minor nucleotides in nucleic acids. MS of macromolecules such as proteins relies on so-called *soft ionization* methods, which allow then ionization of large molecules without degradation.

Two methods are commonly used: *ESI* (*electrospray ionization*) and *MALDI* (*matrix-assisted laser desorption/ionization*). The simplest application of MS is to catalog the *m/e* values for the proteolytic (usually tryptic) digests of a protein. These *m/e* values are then compared with either a database of the predicted tryptic digests of protein or, given a suitable database of protein sequences, with 'virtual digests' performed in real time on a computer. However, this requires that the protein in question already be characterized.

```
Y series

                        <--------------- Y2 --------------->
                                          <-------Y1 ------->
H2NCH(R1H)C(=O)---NHCH(R2H)C(=O)--- NHCH(R3H)C(=O)OH

<---- B1 ---->
<------------- B2 ------------>

B series
```

Fig. 3.3. Illustration of the origin of B and Y series ions in the fragmentation of a hypothetical tripeptide.

For the complete sequencing of an unknown peptide by MS, the protein can be randomly fragmented in a *collision cell* within the mass spectrometer, generating a nested set of fragments. Given a knowledge of the molecular masses of all amino acids, calculating the *m/e* ratios of all these fragments (and more specifically, the differences between them) allows the sequence to be deduced. In reality, the problem is complicated by the fact there are two series of fragments generated at the same time: these are known as the *B series* (N-terminal fragments) and the *Y series* (C-terminal fragments). The problem can be addressed by labeling the protein with, for example, an N-terminal tag, which would add recognized mass to all fragments in the B series. Note that the amino acids leucine and isoleucine cannot be distinguished by MS because they have the same relative molecular mass.

Determination of Protein Structure

X-ray Crystallography

The determination of protein structure by *X-ray crystallography* involves the reconstruction of atomic positions based on the diffraction pattern of *X-rays* through a precisely orientated *protein crystal*. Scattered X-rays cause positive and negative interference, generating an ordered pattern of signals called *reflections*. The amplitude and phase of the scattering depends on the number of electrons in each atom, allowing *electron density maps* to be reconstructed. This requires a mathematical technique called the *Fourier transform*, and the more data used the greater the resolution of the resulting structural model. Structural determination depends on three variables: the *amplitude* and *phase* of the scattering and the *wavelength* of the incident X-rays.

The wavelength of the X-rays is known and the amplitude of the scattering can be calculated from the intensities of the diffraction pattern. The phase of scattering, however, cannot be calculated directly. Phase can be determined by a number of indirect strategies including *multiple isomorphous replacement* (*MIR*) (the use of heavy atom derivatives of the protein to cause anomalous scattering) and *multiwavelength anomalous diffraction* (*MAD*), which uses synchrotron sources to provide X-rays with a range of incident wavelengths. X-ray crystallography used to be a labourious technique, but recent developments in protein expression, robotic workstations that can process hundreds of thousands of crystallization experiments in parallel, and high-throughput MAD analysis allow the relevant data to be collected in only a few weeks.

NMR Spectroscopy

NMR (*nuclear magnetic resonance*) *spectroscopy* is used to determine the structure of small proteins (less than 25 kDa) in solution, and is often applied to proteins that do not readily form crystals. The basis of the technique is that some atoms, including natural isotopes of nitrogen, phosphorus and hydrogen, behave as tiny magnets and can switch between magnetic *spin states* in an applied magnetic field. This is achieved by the absorbance of low wavelength electromagnetic radiation (radio waves), generating *NMR spectra*. The resonance frequency for different atoms is unique, and is also influenced by the surrounding electron density. Thus, the magnetic resonance changes not only for different atoms but also in the context of different chemical groups (this is termed a *chemical shift*), allowing the discrimination between, for example, aryl and aromatic groups. The decay of nuclear magnetic resonance depends on the structure and spatial configuration of the molecule. The *nuclear Overhauser effect* (*NOE*) results from the transfer of magnetic energy through space if inter. acting nuclei are less than 0.5 nm apart and results in the appearance of symmetrical peaks superimposed over the normal one-dimensional NMR spectrum. The analysis of NMR spectra thus allows a set of *distance constraints* to be built up, which can be used to determine the three-dimensional spatial arrangement of atoms.

Other Methods

X-ray crystallography and NMR spectroscopy account for the vast majority of three-dimensional structural data available for proteins. However, some proteins are both difficult to crystallize and too large for NMR. In such cases, other spectroscopic and related methods are required. For large integral membrane proteins, which are generally difficult to crystallize, it is often necessary to study the membrane as a whole. A modification of NMR—*magic angle spinning NMR*—allows sharp NMR signals to be obtained from preparations in films. The regular spacing of proteins in membranes also permits alternative techniques to be used, such as *electron microscopy*.

The electron micrograph is used as a diffraction grating. In this way, structural information can be derived but, to date, with a lower resolution than other methods. Alternatively, the membrane with its proteins in place can be used as a two-dimensional crystal, and diffraction patterns can be obtained from X-rays with a low angle of incidence on the preparation. This method can also be used to determine the structure of fibrous proteins such as collagen and keratin, and is known as *fiber diffraction*. A useful complement to X-ray crystallography and NMR spectroscopy is *circular dichroism (CD) spectroscopy*, which exploits the optical activity of asymmetric molecules as shown by their differing absorption spectra in left and right circularly polarized light. CD spectroscopy between 160 and 240 nm can provide information about protein secondary structure, since proteins containing predominantly α-helices and proteins consisting mainly of β-sheets generate distinct spectra.

Gene and Protein Expression Data

Global Expression Analysis

Gene expression can be studied at either the RNA or protein level. Traditionally, expression has been studied on a gene-by-gene basis using techniques such as Northern and Western blots. More recently, methods have been developed for *global expression analysis*, that is, the study of all ganes simultaneously. Such studies generate very large amounts of data, which must be mined for relevant information using bioinformatics tools, but what forms does this data take? For analysis at the RNA level, *direct sequence sampling* from RNA populations or cDNA libraries, or even from sequence databases, can be useful.

A simple approach is to sequence say 5000 randomly picked clones from a cDNA library. Abundant mRNAs would appear at a higher frequency among the sampled sequences than rare ones, and statistical analysis of these data would allow relative expression levels to be determined. Although simple in concept, this type of experiment is expensive because of all the sequencing reactions that must be carried out. A more sophisticated technique is *serial analysis of gene expression* (*SAGE*). In this method, very short sequence tags (usually 8-15 nt) are generated from each cDNA and hundreds of these are joined together to: form a concatemer prior to sequencing. Thus, in one sequencing reaction, information on the abundance of hundreds of mRNAs can be gathered.

Each *SAGE tag* uniquely identifies a particular gene, and by counting the tags, the relative expression level of each gene can be determined. The advantage of all direct' sampling methods is that expression levels are automatically converted into numbers, that is, the data are digital. Although direct sampling is a powerful approach to expression analysis, it is labour intensive. The technology that has had the widest impact on global RNA expression profiling is hybridization to *DNA arrays* (*DNA chips*).

As discussed in more detail below, the principle advantage of DNA arrays is that the expression levels of all genes can be monitored in parallel in a single experiment Expression data are obtained as signal intensities, and these are clustered to identify similarly expressed genes. Array-based assay formats have also been developed for proteins but these 'protein chips' are not yet as versatile as the equivalent DNA chips. Currently" the most widely used technology for protein expression analysis is separation by *two-dimensional polyacrylamide gel electrophoresis* (*2D-PAGE*), generating a unique pattern of dots (each dot representing a single protein). In 2D-PAGE experiments, protein expression data are obtained as signal intensities for each of the dots. However, unlike microarray experiments where each signal can be linked to its corresponding gene, each protein signal on a 2D gel has to be individually annotated by *mass spectrometry*.

DNA Microarrays

Nucleic acids (DNA and RNA) can form double-stranded molecules by *hybridization*, that is, complementary base pairing. The specificity of nucleic acid hybridization is such that a particular DNA or RNA molecule can be labeled (e.g. with a radioactive or fluorescent tag) to generate a *probe*, and can be used to isolate a complementary molecule from a very complex mixture, such as whole genomic DNA or whole cellular RNA. This specificity also allows thousands of hybridization reactions to be carried out simultaneously in the same experiment. A *DNA microarray* or *DNA chip* is a dense grid of DNA elements (often called *features* or *cells*) arranged on a miniature support, such as a nylon filter or glass slide. Each feature represents a different gene.

The array is usually hybridized with a *complex RNA probe*, that is, a probe generated by labeling a complex mixture of RNA molecules derived from a particular cell type. The composition of such a probe reflects the levels of individual RNA molecules in its source. If nonsaturating hybridization is carried out, the intensity of the signal for each feature on the microarray represents the level of the corresponding RNA in the probe, thus allowing the relative expression levels of thousands of genes to be visualized simultaneously. Two major technologies are used in microarray manufacture. The cheaper and currently more widely accessible method involves the robotic spotting of individual DNA clones onto a coated glass slide. Such *spotted DNA arrays* can have a density of up to 5000 features per square cm. The features comprise double-stranded DNA molecules (genomic clones or cDNAs) up to 400 bp in length and must be denatured prior to hybridization. Prefabricated arrays can be purchased from a number of companies and many laboratories have their own in-house production facilities.

The alternative method is *on-chip photolithographic synthesis*, in which short oligonucleotides are synthesized *in situ* during chip manufacture. These arrays are known as *GeneChips* and are manufactured exclusively by the US company Affymetrix Inc. They have a density of up to 1000000 features per square cm, each feature comprising up to 10^9 singlestranded oligonucleo-tides 25 nt in length. Due to the reduced hybridization specificity of a short oligonucleotide, each gene on a GeneChip is represented by 20 features (20 nonoverlapping oligos), and 20 mismatching controls are included to normalize for nonspecific hybridization. Both technologies can be used to look at differential gene expression between samples (e.g. healthy tissue vs. disease tissue).

Fluorescent probes are used for spotted DNA arrays, since different fluorophores can be used to label different RNA populations. These can be simultaneously hybridized to the same array, allowing differential gene expression to be-monitored directly. Typically, Cy3 is used to label one probe and Cy5 to label

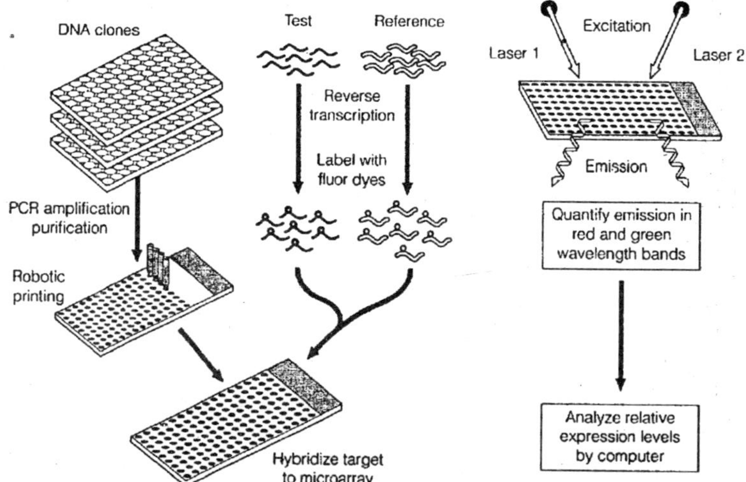

Fig. 3.4. The process of differential expression measurement using a DNA microarray.

the other. Cy3 fluoresces bright red and Cy5 fluoresces bright green. If a particular RNA is present only in the Cy3labeled probe, the corresponding feature on the array appears red.

If another RNA is present only in the Cy5-labeled probe, that spot appears green. RNAs found in both probes would hybridize in equivalent amounts, and these features would appear yellow. Dual labeling is not used on GeneChips. Instead, hybridization is carried out with separate probes on two identical chips and the signal intensities are measured and compared by the accompanying analysis software.

2D Protein Gels

2D-PAGE is a well-established biochemical technique in which proteins are separated on the basis of two separate properties: their *isoelectric point* (*pI*) and their *molecular mass*. Separation in the first dimension is nowadays carried out by isoelectric focusing in an immobilized pH gradient. The pH gradient is generated by a series of buffers, and an *immobilized pH gradient* is produced by covalently linking the buffering groups to the gel, therefore preventing migration of the buffer itself during electrophoresis. *Isoelectric focusing* means allowing proteins to migrate in an electric field until the pH of the buffer is the same as the pI of the protein. The pI of the protein is the pH at which it carries no net charge and therefore does not move in the applied electric field. Next, the gel is equilibrated in the detergent *sodium dodecylsulfate (SDS)*, which binds uniformly to all proteins and confers a net negative charge. Therefore, separation in the second dimension can be carried out on the basis of molecular mass.

After the second-dimension separation, the protein gel is stained with a universal dye to reveal the position of all protein spots. Reproducible separations can then be carried out with similar samples to allow comparison of protein expression levels. Each protein can be characterized using x,y co-ordinates, where x corresponds to the pI and y to the molecular mass. Differential expression may be shown by the presence/absence of particular spots or by different spot intensities between samples. However, one complication that must be taken into account is that different forms of *post-translational protein modification* (e.g. phosphorylation, glycosylation) also contribute to position shifts and quantitative variations on 2D gels.

Protein Interaction Data

Importance of Protein Interactions

Most of the functions of a cell are carried out by proteins. These functions depend on *protein interactions*, that is, proteins interacting with each other and with other types of molecule. The disruption of protein interactions (by mutation, infection with a pathogenic organism, drug and chemical abuse, etc.) can have a variety of ill effects. Conversely, therapeutic drugs work by interacting beneficially with proteins. It is useful to know about *protein-protein interactions* for several reasons. Many proteins either exist in permanent *multi-subunit complexes* (e.g. hemoglobin) or assemble into complexes in order to function (e.g. DNA polymerase, RNA polymerase).

The functions of many other proteins are influenced by the formation of *transient complexes*. This includes many signaling proteins, and proteins involved in the regulation of processes such as the cell cycle and membrane transport. Through their interactions, proteins can influence each other's activity. This may be by binding (e.g. proliferating cell nuclear antigen [PCNA] stimulates the activity of DNA polymerase δ) or by catalysis (e.g. protein kinases dephosphorylate residues on other proteins and modulate their activities). The fact that the human genome appears to contain only 35000-50000 protein coding genes, compared with 4000 for the single *Escherichia coli* cell, indicates that organism complexity is achieved by means other than total gene number. The ability of proteins to interact in concert through diverse complexes and pathways is one way in which complexity can be achieved. In addition to protein-protein interactions, it is also useful to know about the interactions between proteins and other molecules. *Protein-nucleic acid interactions,* for example, are central to a number of critical cellular processes (replication, recombination, transcription, RNA processing and protein synthesis, to name but a few).

Interactions between proteins and small molecules are also important. This underlies many signaling processes (e.g. receptor-ligand binding) and often controls protein activity. For example, the activity of the *E. coli* TrpR repressor, which blocks transcription of genes involved in tryptophan biosynthesis, is regulated by the binding of tryptophan itself. Drug development centers on the understanding and exploitation of protein-small ligand interactions. In this topic, we briefly discuss some of the many different methods that can be used to characterize protein interactions. Bioinformatics is assuming an important role in this field as large amounts of interaction data accumulate. It is necessary to filter the data, extracting the reliable and meaningful interactions, build databases of interacting proteins, and use this information to reconstruct molecular pathways and networks.

Genetic Methods

In genetically amenable species such as bacteria and yeast, protein-protein interactions can be inferred from *genetic analysis*. The basis of this type of approach is that interactions between two given proteins, X and Y, can be studied by looking at the behaviour of mutations in their corresponding genes, X and Y. For example, a *suppressor mutation* restores the phenotype of a given primary mutation to normal or near normal. One explanation is that a primary mutation, in gene X, causes a conformational change in protein X that prevents its interaction with protein Y, therefore causing a loss of function.

The suppressor mutation, in gene Y, introduces a complementary change in protein Y that restores the interaction and thus rescues the mutant phenotype. Another genetic test for protein interactions is the *synthetic lethal screen*. In this case, single mutations in genes X and Y in different individuals are nonlethal because the conformational changes in the corresponding proteins are minor and interaction between proteins X and Y is preserved. However, by combining the mutations in the same individual, the interaction between X and Y is disrupted and a lethal phenotype is observed. The formation of a multi-subunit complex can be demonstrated using *dominant negative* mutants. In this case, a nonfunctional mutant form of the protein is produced in excess, for example by injecting mRNA into the cell or introducing a recombinant gene that is expressed at a high level. The mutant form the protein sequesters the normal, active form into nonfunctional complexes resulting in a loss of function phenotype.

Affinity Methods

Genetic methods infer protein interactions but do not demonstrate them directly. In contrast, there are several physical methods that provide direct evidence for protein interactions by exploiting the *affinity* (tendency to bind specifically) of proteins for each other. Typically, one molecule is immobilized, and used as a bait to trap interacting proteins, which can then be purified and characterized. *Affinity chromatography* involves attachment of the bait molecule to a matrix, such as a Sepharose column. Protein extracts are then passed through the column and washed through with a low-salt buffer. Proteins that interact with the bait are retained, while others are eluted with the buffer. The bound proteins can themselves be eluted using a higher concentration of salt in thebuffer. By using different salt concentrations, it is possible to discriminate between proteins that bind weakly and those that bind strongly to the bait. One popular variation of this method is known as *GST-pulldown*.

In this technique, the bait protein is expressed as a fusion with the enzyme *glutathione-S-tranferase*. This enzyme has a very high affinity for its substrate, *glutathione*, so the fusion protein will bind very strongly to glutathione-coated Sepharose beads. Proteins that interact with the GST-fusion protein will be retained on the chromatography column while noninteracting proteins are eluted. The advantage of this method is that any protein can be used as a bait, provided it can be expressed as a GST fusion. However, it is important to run a control chromatography experiment using GST alone, to weed out those proteins that interact with GST itself! Another affinity method, *co-immuno-precipitation*, involves the precipitation of protein complexes using an antibody specific for one component of the complex. A cell lysate is prepared under conditions that preserve protein-protein interactions. An antibody raised against protein X is then mixed with the lysate, resulting in the precipitation of protein X and any proteins interacting with it.

One problem with both affinity chromatography and co-immunoprecipitation is that these methods can isolate proteins that interact indirectly as well as those that interact directly with the bait. *Cross-linking* can be used to identify direct interactions and to probe the architecture of protein complexes. Recent technological advances, led by the company Ciphergen, have allowed affinity assay formats to be extensively miniaturized, culminating in the development of *protein chips* (also called *protein microarrays* because they are similar in principle to the DNA microarrays). Protein chips are manufactured in the same manner as DNA arrays, that is, by the robotic spotting of small amounts of liquid onto a solid substrate such as a glass slide. Capture agents may be broad in their scope (e.g. surface chemistries that capture whole classes of related proteins) or narrow and specific (e.g. antibodies).

Molecular and Atomic Methods

Protein-protein interactions and protein interactions with other molecules can be characterized using a number of molecular and atomic methods. For example, *X-ray crystallography* and *nuclear magnetic resonance spectroscopy* can be used to determine the structure of proteins either as uncomplexed molecules or in complexes with ligands, cofactors and substrates. Another useful technique is *fluorescent resonance energy transfer* (*FRET*), in which interaction between protein X and protein Y is shown by the transfer of energy from an excited donor fluorophore (conjugated to one protein) to an acceptor fluorophore (conjugated to the other).

FRET occurs only when the two fluorophores are up to 10 nm apart, and can be detected by a change in the emission wavelength of the acceptor fluorophore. Instead of conjugated fluorophores, the interacting proteins may be fused to naturally occurring bioluminescent proteins (such as *green fluorescent protein, GFP*) in which case the technique may be termed *bio-luminescent resonance energy transfer* (*BRET*). The advantage of FRET/BRET is that it detects transient as well as stable interactions. *Mass spectrometry* (*MS*) can also be used to study protein-protein interactions. This type of experiment is usually carried out to determine the composition of a multi-subunit protein complex, and has been used for example to characterize the yeast spliceosome and anaphase-promoting complex.

The procedure is to isolate an intact complex and subject it to matrixassisted laser desorption/ionization (MALDI) MS to annotate individual proteins. This can be done either with or without prior electrophoretic separation of the proteins. A final group of molecular methods is worth mentioning because the technologies are exploited on protein chips and can be integrated directly with MS analysis. One such method is *surface plasmon resonance* (*SPR*) *spectroscopy*, which is used on chips marketed by the US biotechnoloy company BIAcore. *Surface plasmon resonance* is an optical resonance phenomenon occurring when *plasmon waves* (charge density oscillations) become excited at the interface between a metal surface and a liquid. Interaction between protein X, which is immobilized on the metal surface, and protein Y which is free in solution, is revealed by a change in the refractive index of the surface layer.

Another method is *surface-enhanced laser desorption/ionization* (*SELDI*). In this method, interacting proteins are trapped by affinity to a capture agent, such as an antibody, which is immobilized on a solid substrate. After washing and optional tryptic digestion of the interacting proteins, the substrate is coated with matrix compound and used directly for MALDI MS analysis. The advantage of SELDI over standard MALDI MS is that it provides more uniform mass spectra and allows protein quantification.

Library-based Methods

All of the above methods are very powerful for detecting and characterizing protein interactions. However, their universal disadvantage is that only a few samples can be processed at anyone time and the proteins identified as interactors cannot conveniently be associated with the genes that encode them. *Library methods* are based on traditional DNA expression libraries and solve both these problems by using a highly parallel assay format in which proteins are expressed from their associated cDNAs. Standard *expression libraries* and *phage display libraries* can be used for interaction analysis but in both cases the interactions occur *in vitro,* where the protein folding and recognition conditions do not reflect those in the cell. Undoubtedly the method which has had the greatest impact on protein-protein interaction research is the *yeast two-hybrid* (*Y2H*) *system* and related technologies, in which interacting proteins are used to assemble a functional transcription factor.

Protein X is expressed as a fusion (a hybrid) with the DNA binding domain of a transcription factor, to generate a *bait*. An expression library is then generated in which each clone is expressed as a fusion protein with the transcription factor's transactivation domain; these hybrids are potential *prey*. The final component of the system is a reporter gene that is activated specifically by the two-hybrid transcription factor. Mating between haploid yeast cells carrying the bait construct and those carrying the library of prey results in diploid cells carrying both components. In those cells where the bait interacts with the prey (protein Y) the transcription factor is assembled and the reporter gene activated, allowing the cells to be isolated and the DNA sequence of the interactor identified. Y2H screens have been used to search for candidate interactions in a variety of molecular systems, including the entire yeast genome, several virus genomes and in specific sets of proteins from the fruit fly and the nematode worm. Very large amounts of data are generated in such screens. The main problem with this type of assay, and the reason that bioinformatics is so important, is the relatively high level of false-positive and false-negative results, meaning that accessory information has to be included to determine which interactions are genuine and which are spurious.

Databases-contents, Structure and Annotation

File Formats

Three Common Sequence Formats

If biological data is to be used by computer programs, it must be presented in a standard format that can be read by computer. It is very common to put data in text files. As the name suggests, these files contain text that can be read by a human being as well as a computer. They are rather like the files used by word-processing packages to hold documents, but there is one important difference: text files hold (almost) only the text and little auxiliary information about formatting.

Many bioinformatic databases and software applications are designed to work with sequence data, and this requires a standard format for inputting nucleic acid and protein sequence information. Three of the most common *sequence formats* are NBRF/PIR (National Biomedical Research Foundation/Protein Information Resource), FASTA and GDE. Each of these formats has facilities not only for representing the sequence itself, but also for inserting a unique *code* to identify the sequence and for making *comments* which may include for example the name of the sequence, the species from which it was derived, and an accession number for GenBank or another appropriate database.

NBRF/PIR format

Note that the first line begins with '>*P1;*' which specifies a protein sequence. If this was a nucleic acid sequence, it would begin with '>*N1;*'. The semicolon is followed by a code, in this case '5H1B_CAVPO', which is a unique sequence identifier. Serotonin is also known as 5-hydroxytryptamine, thus 5HIB identifies the protein as serotonin receptor IB, while CAVPO identifies its source as the guinea-pig (*Cavia porcellus*). There follows a *comment line,* and the rules allow this line to be of more or less any length so it can either be empty or far too wide to fit on a printed page. Then the sequence itself follows and is terminated by an asterisk (*). It is conventional to give files in this format the extension '.pir' or '.seq'.

FASTA format

The first line begins with '>' but there is no designation of protein or nucleic acid sequence. The code is entered next and this is followed (on the same line) by comments, although it is conventional to delimit the comments with a '|' symbol. As with the NBRF/PIR format there is no limit to the length of the first line. One point to note about FASTA files is that they allow lower-case letters for the amino acids. Files in this format commonly have the extension '.fasta'.

GDE format

This is essentially the same as the FASTA format, but the '>' symbol in the first line is replaced by '%'. Files in this format have the extension '.gde'. All three file formats *ignore spaces* and *carriage returns*. This allows sequences to be typed out in a manner that is convenient for the user. For example, a space has been

(a)

```
>P1;5H1B_CAVPO
Guinea pig serotonin receptor accession: O08892
MGNPEASCTP PAVLGSQTGL PHANVSAPPN NCSAPSHIYQ DSIALPWKVL LVVLLALITL
ATTLSNAFVI ATVYRTRKLH TPANYLIASL AFTDLLVSIL VMPISTMYTV TGRWTLGQAL
CDFWLSSDIT CCTASIMHLC VIALDRYWAI TDAVGYSAKR TPRRAAGMIA LVWVFSICIS
LPPFFWRQAK AEEEVLDCLV NTDHVLYTVY STGGAFYLPT LLLIALYGRI YVEARSRILK
QTPNKTGKRL TRAQLITDSP GSTSSVTSIN SRAPEVPCDS GSPVYVNQVK VRVSDALLEK
KKLMAARERK ATKTLGVILG AFIVCWLPFF IISLVMPICK DACWFHMAIF DFFTWLGYLN
SLINPIIYTM SNEDFKQAFH KLIRFKCTT
*
```

(b)

```
> 5H1B_CAVPO  O08892|guinea pig serotonin receptor
MGNPEASCTP PAVLGSQTGL PHANVSAPPN NCSAPSHIYQ DSIALPWKVL LVVLLALITL
ATTLSNAFVI ATVYRTRKLH TPANYLIASL AFTDLLVSIL VMPISTMYTV TGRWTLGQAL
CDFWLSSDIT CCTASIMHLC VIALDRYWAI TDAVGYSAKR TPRRAAGMIA LVWVFSICIS
LPPFFWRQAK AEEEVLDCLV NTDHVLYTVY STGGAFYLPT LLLIALYGRI YVEARSRILK
QTPNKTGKRL TRAQLITDSP GSTSSVTSIN SRAPEVPCDS GSPVYVNQVK VRVSDALLEK
KKLMAARERK ATKTLGVILG AFIVCWLPFF IISLVMPICK DACWFHMAIF DFFTWLGYLN
SLINPIIYTM SNEDFKQAFH KLIRFKCTT
```

(c)

```
%5H1B_CAVPO  O08892|guinea pig serotonin receptor
MGNPEASCTP PAVLGSQTGL PHANVSAPPN NCSAPSHIYQ DSIALPWKVL LVVLLALITL
ATTLSNAFVI ATVYRTRKLH TPANYLIASL AFTDLLVSIL VMPISTMYTV TGRWTLGQAL
CDFWLSSDIT CCTASIMHLC VIALDRYWAI TDAVGYSAKR TPRRAAGMIA LVWVFSICIS
LPPFFWRQAK AEEEVLDCLV NTDHVLYTVY STGGAFYLPT LLLIALYGRI YVEARSRILK
QTPNKTGKRL TRAQLITDSP GSTSSVTSIN SRAPEVPCDS GSPVYVNQVK VRVSDALLEK
KKLMAARERK ATKTLGVILG AFIVCWLPFF IISLVMPICK DACWFHMAIF DFFTWLGYLN
SLINPIIYTM SNEDFKQAFH KLIRFKCTT
```

Fig. 3.5. The sequence of a guinea-pig serotonin receptor in (a) NBRF/PIR format; (b) FASTA format; and (c) GDE format.

inserted every 10 amino acid residues and a carriage return after every 60, making it much easier to manually count the residues and identify amino acids at specific positions in the sequence. Note, however, that most standard word-processing software packages do not ignore blank spaces. For some purposes it may be desirable or necessary to construct files from unpublished and preliminary sequence data, and if programs such as Microsoft Word or Corel WordPerfect are used, the results can be unpredictable. If using Word, use text only mode with a non-proportional font or, preferably, use a simple text editor such as Notepad.

To illustrate this point, consider the creation of the following very simple NBRF/PIR file:

```
>P1;MY_CODE
my cat
MYCATSATINMYLAP*
```

Despite the fact that this protein is clearly fictitious, the format is perfectly correct and it should be possible to search for the peptide sequence in other proteins. However, by typing this sequence into Microsoft Word and saving it as a Word document (cat.doc), the file proves to be over 19 thousand bytes in length, and therefore obviously contains much more than the simple text. By saving the file as *text with line breaks* (cat.txt), the file size is reduced to 39 bytes, which seems more reasonable. However, inspection of the contents of cat.txt reveals two extra characters at the end of each line and another at the very end of the file. It is therefore best to avoid word processors and use text editors for the preparation of sequences. If a word processor is used, the file should be saved as text and sent by ftp as ASCII, and a text editor should then be used on the computer where the sequence analysis is carried out, to check the integrity of the file. Another point to bear in mind is that the first line of a FASTA file and the second line of an NBRF/PIR file might be extremely long and it is essential not to cut it up by inserting carriage returns, otherwise the comments might be read as part of the sequence.

Files for Aligned Sequences

The output from *sequence-alignment programs* can be in anyone of a number of formats. All three formats discussed above are suitable for dealing with aligned sequences but there are several formats designed specifically for alignment output. In order to achieve the alignments, *gaps* must be introduced and these are represented either by hyphens or dots. *Multiple sequence format* (*MSF*) is used by several software tools. *PHYLIP* (*phylogenetic inference package*) is the output format of the software of that name and CLUSTALW /X (FI) has its own *ALN format.*

Files of Structural Data

The raw materials for bioinformatic studies on macromolecular structures are *PDB files*. These are text files using a format devised by the Protein Data Bank. Such files contain orthogonal atomic co-ordinates together with annotations, comments and experimental details. The most important aspect of PDB files is that the' ATOM' lines are laid out in columns of characters *not* columns of words. Compare the first ATOM lines:

ATOM	1	N	VAL	16	29.582	19.112	38.968
ATOM	1	N	ILE E	16	–9.947	23.613	20.817
ATOM	1	N	ALA	1	14.702	–10.824	3.425

The last three columns show the orthogonal co-ordinates (*x*, *y*, z) of the atom and they are deduced by counting the positions including the spaces, *not* by counting the words. This is because the *x* co-ordinate is the sixth word in the first and third cases but the seventh word in the second case, because a new word, the chain identifier (E), has been inserted. Other points that emerge are: the atoms are numbered consecutively; amino acids are represented by three letters; the ends of the lines may or may not contain line numbers; NMR files do not have a *resolution* REMARK; and NMR files typically contains several models corresponding to different conformations.

Annotated Sequence Databases

Primary Sequence Databases

The *primary sequence databases* are repositories for raw sequence data, and can be accessed freely over the World Wide Web (WWW). There are three such databases, comprising the *International Nucleotide Sequence Database Collaboration*. These are *GenBank*, maintained by the *National Center for Biotechnology Information* (*NCBI*), the *Nucleotide Sequence Database* maintained by the *European Molecular Biology Laboratory* (*EMBL*), and the DNA *Databank of Japan* (*DDBJ*). New sequences can be deposited in any of the databases since they exchange data on a daily basis.

The databases contain not only sequences but also extensive annotations. Much of the introductory part of the file is self-explanatory, containing information such as the locus name, the accession number, the source species literature references and the date of submission. An important section of the file is the *features table*, which describes interesting features of the sequence. Since GenBank is a nucleic acid repository, the fact that there is a protein-coding region is a feature in the entry. Note that *BTEB* is a very simple gene that has no introns. If there were introns, the CDS (coding sequence) feature would be more complicated: the entry would be extended to indicate the base positions of the exons, delimited by commas. For example, if there was a second exon encoding a further 20 amino acids residues, the CDS feature would read as follows: 1265..1999,2100..2159.

Subsidiary Sequence Databases

The main sequence databases have a number of subsidiaries for the storage of particular types of sequence data. For example, *dbEST* is a division of GenBank which is used to store *expressed sequence tags* (*ESTs*). Other divisions of GenBank include *dbGSS*, which is used to store single-pass *genomic survey sequences* (*GSSs*), *dbSTS*, which is used to store *sequence tagged sites* (*STSs;* unique genomic sequences that can be used as physical markers) and the *HTG* (*high-throughput genomic*) *division,* which is used to store unfinished genomic sequence data.

Submission of Sequences

The robustness of data submitted to the primary sequence databases is important in the context of bioinformatics software. Clearly, the integrity of the scientists who submit the data is not readily checked by computers but errors must be avoided in database consistency. It is essential that the data are submitted in a supported format and that the submission is carried out by means of software provided by the database curators. Examples are *WebIn* provided by EMBL (www.ebi.ac.uk/embl/Submission) and *BankIt* provided by the NCB! (http://www.ncbi.nlm.nih.gov/Banklt/). each of which can be used to submit sequences to the databases over the WWW. A powerful stand-alone software tool, *Sequin,* is provided by the NCBI and can be used on UNIX, PC/Windows and Macintosh systems for sequence submission for those with no WWW access (http://www.ncbi.nlm.nih.gov/Sequin/index.html).

SWISS-PROT and TrEMBL

SWISS-PROT and the related database *TrEMBL* (Translated EMBL) are repositories for annotated protein sequences. The entry contains large numbers of annotations, including a *features table* before the sequence. Each line begins with two letters, many of which are self-explanatory, for example ID (identity), AC (accession number), DT (date), DE (description), GN (gene name), CC (comment). Continuation lines are indicated by the symbols -!- at the start of a section and indents thereafter. Characteristic features of SWISS-PROT entries include the DR (reference), KW (key words) and FT (features) fields. It is the presence of these careful and extensive annotations that makes SWISS-PROT so popular with biochemists. For example, there is a fairly comprehensive description of the protein and its function but also (in the DT field) cross-references to the relevant entries in the secondary databases PROSITE, PRINTS and Pfam.

SWISS-PROT provides the most up-to-date and extensively annotated information on protein sequences and its quality reflects its active management by human curators. TrEMBL (translated EMBL) is another database in the same format. The entries in TrEMBL are derived from translation of all coding sequences in the EMBL Nucleotide Sequence Database that are not already in SWISS-PROT. As further data ensure the reliability of annotations, TrEMBL entries are moved to SWISS-PROT.

Database Interrogation

Detailed queries of the text annotation in the databases discussed above can be carried out using tools like SRS and Entrez. However, a comparison shows how a user can cross-reference between these databases. Let us assume, for example, that searching GenBank with a new sequence obtained in the laboratory identifies the gene as particularly interesting for a research project. How do we find the corresponding entry in SWISS-PROT? Note that the IDs are different, and, although SWISS-PROT has alternative GN entries (gene names), they do not correspond to the name on the GenBank file. The way to find the correct SWISS-PROT file is to search the SWISS-PROT database for the accession number D31716. Although SWISS-PROT has its own accession number, D31716 can be found as a DR field entry. The SWISS-PROT site or one of its mirrors will successfully locate the entry with D31716 as the search string.

Genome and Organism-specific Databases

Organism-specific Resources

The annotated sequence databases are general to all organisms, and contain data relevant to viruses, bacteria, microbial eukaryotes, animals and plants, as well as recombinant molecules produced in the laboratory. However, there are also many databases devoted to particular organisms, and their numbers are increasing as further genome projects are initiated. Typically, such databases contain not only sequence data but also information on gene expression, mutant phenotypes, genome maps, genome sequencing projects and relevant scientific literature, and provide links to resources for obtaining clones, mutants and for contacting researchers. A selected list of organism-specific databases is provided in Table 3.1, but this represents only a small fraction of the resources available.

Table 3.1. A small selection of organism-specific genomic databases available on the WWW.

Organism	*Database/resource*	*URL*
Escherichia coli	EcoGene	http://bmb.med.miami.edu/EcoGene/EcoWeb/
	EcoCyc (Encyclopedia of *E. coli* genes and metabolism)	http://ecocyc. pangeasystems.com/ecocyc/ecocyc.html
	Colibri	http://genolist.pasteur.fr/Colibri/
Bacillus subtilis	SubtiList	http://genolist.pasteur.fr/SubtiList/
Saccharomyces cerevisiae	*Saccharomyces* Genome Database (SGD)	http://genome-www. stanford.edu/Saccharomyces/
Plasmodium falciparum	PlasmoDB	http://PlasmoDB.org
Arabidopsis thaliana	MIPS *Arabidopsis* thaliana Database (MAtDB)	http://mips.gsf.de/proj/thal/db
	The *Arabidopsis* information resource (TAIR)	http://www.arabidopsis.org/
Drosophila melanogaster	FlyBase	http://flybase.bio.indiana.edu/
Caenorhabditis elegans	A C. *elegans* DataBase (ACeDB)	http://www.acedb. org/
Mouse	Mouse Genome Database (MGD)	http://www.informatics.jax.org/
Human	Online Mendelian Inheritance in Man (OMIM)	http://www.ncbi.nlm.nih.gov/omim

The user interested in an organism that is not listed in Table 3.1 could try using a search engine to find a useful resource, but there are also a number of excellent gateways available on the WWW, which provide information on multiple organism-specific resources and links to the relevant sites. A number of these gateways are listed in Table 3.2.

Database Formats

Genomic databases need to facilitate the storage and analysis of large amounts of data, but must also have a user-friendly front-end graphical display to allow relevant data to be displayed and analyzed. A number of viewers and browsers have been developed for genomic sequence analysis and annotation. These include Artemis, Apollo, EnsEMBL and GoldenPath. One of the most versatile database formats is *ACeDB*. This was originally designed for research on the nematode worm *Caenorhabditis elegans* (*A C. elegans DataBase*). ACeDB has an object

Table 3.2. Useful gateway sites providing information and links to multiple, organism-specific and genomic resources.

Gateway site	*URL*
NCBI Genomic Biology	http://www.ncbi.nlm.nih.gov/Genomes/index.html
GOLD (Genomes Online Database)	http://wit.integratedgenomics.com/GOLD/
Organism-specific genome databases	http://www.unl.edu/stc-95/ResTools/biotools/biotools10.html
TIGR Microbial Database	http://www.tigr.org/tdb/mdb/mdbcomplete.html
Bacterial genomes	http://genolist.pasteur.fr/
Yeast databases	http://genome-www.stanford.edu/Saccharomyces/yeast_info.html
EnsEMBL genome database project	http://www.ensembl.org/
MIPS (Munich Information Center for Protein Sequences)	http://mips.gsf.de

orientated database architecture rather than a simple collection of data and it has been used to handle data on other organisms, including the yeast *Scizosaccharomyces pombe* and humans.

ACeDB has a graphical user interface with displays and tools designed for genomic data. Other features of ACeDB include *AQL* (*ACeDB query language*), interfaces with Perl and Java, WWW interfaces (of which AceBrowser is the current and supported version), WinAce (the Windows95/NT version of ACeDB), CITA (a CORBA interface to the database) and Acembly (a sequence assembly system). The URL is http://www.acedb.org/.

Table 3.3. Database tools for displaying and annotating genomic sequence data.

Viewer format	*URL for further information and tutorials*
Artemis	http://www.sanger.ac.uk/Software/Artemis
ACeDB	http://www.acedb.org/Tutorial/brief-tutorial.shtml
Apollo	http://www.ensembl.org/apollo/
EnsEMBL	http://www.ensembl.org
NCBI map viewer	http://www.ncbi.nlm.nih.gov/
GoldenPath	http://genome.ucsc.edu/

Finding Organism-specific Databases

This section provides a number of worked examples of how to find organism-specific databases, resources and information.

A good starting point for this type of search is the Genomes Online Database (GOLD). Once the GOLD top page·has been accessed (http://witintegratedgenomics.com/GOLD/), the site can be searched for information on any organism, for example Ebola virus, the bacterium *Escherichia coli*, the fruit fly *Drosophila melanogaster*, and humans.

Ebola virus

Information on Ebola virus can be found by clicking on the *Viruses* hyperlink under 'Other Links' on the GOLD top page. This accesses the European Bioinformatics Institute page of completed viral genomes, which currently lists nearly 700 viruses whose genomes have been sequenced, together with the corresponding European Molecular Biology Laboratory Nucleotide Sequence Database file, the sequence in FAST A format and the sequence retrieval system entry. The URL is http:/ /www.ebLac.uk /cgi-bin/ genomes/genomes.cgi?genomes=viruses.

Escherichia coli

Entering *Escherichia coli* as a search term after accessing the SEARCH GOLD query form pulls records on nine different bacterial strains. Resources are listed in a table with the following headings: *Organism, Tree, Information, Size/ORF-number, Data Search, Institution, Funding, Genome Database, Status* and *Publication*. Of these, the links provided under the headings *Information* and *Genome Database* are the most useful. There are more than 20 resources listed including some general ones [e.g. National Center for Biotechnology Information (NCBD, SWISS-PROT, TIGR (The Institute for Genomic Research) and some specific ones (e.g. EcoCyc, EcoGene, Colibri). The E. *coli* resources can be accessed directly via hyperlinks, and many of the sites have lists of further resources. For example, the EcoCyc page, the Encyclopedia of E. *coli* genes and metabolism, contains a link to 'other information on E. *coli*' which is a page containing 13 further links to E. *coli* resources on the WWW. Similarly, the Colibri site has a link 'other sites related to E. *coli*' which lists over 20 additional resources.

Drosophila melanogaster

When *Drosophila melanogaster* is used as a search term, the *Genome Database* links provided by GOLD are as follows: NCBI, BDGP (Berkeley *Drosophila* Genome Project), FlyBase-UK, FlyBase-USA, EBI-Proteome, KEGG and IBM- Annotation. Under *Information,* a taxonomy resource and another comprehensive *Drosophila* site, the Interactive Fly, are also listed. FlyBase-UK links to http:// www.edgp.ebLac.uk/ and provides a large number of *Drosophila* resources, including clone orders, annotated sequences, raw sequence data in FASTA format, sequence sets and access to sequence analysis tools.

Human

When *Homo sapiens* is used as a search term, links provided by GOLD are as follows: NCBI, ORNL, RIKEN, EnsEMBL, Proteome-EBI, Sanger Centre HOWDY and IBM-Annotation. Clicking on NCBI links to the NCBI human genome resources page, which contains images of human chromosomes. The user can select from a drop-down menu of clones, genes, physical maps, genetic maps and variation and then click on any of the chromosomes to see the information available for the chromosome chosen under that heading. The NCBI server also provides access to UniGene, Online Mendelian Inheritance in Man (OMIM) and other miscellaneous databases.

Miscellaneous Databases

Database Resources

Databases are essentially large storage devices for scientific and other data. They can be searched and cross-referenced either over the Internet or using downloaded versions on local computers or computer networks. Specific types of database are discussed in different topics throughout this book. For example, the three primary nucleic acid databases [GenBank, the European Molecular Biology Laboratory (EMBL) Nucleotide Sequence Database and the DNA Databank of Japan] are discussed already. These are called *primary databases* because they store raw sequence data. Similarly, SWISS-PROT and TrEMBL are the major primary databases for the storage of protein sequences.

There are also secondary databases of protein families and sequence patterns such as PROSITE, PRINTS and BLOCKS. These are called *secondary databases* because the sequences they contain are not raw data, but have been derived from the data in the primary databases. There are also many organism-specific databases containing information, links and resources dedicated to particular species. In this topic, we discuss some of the remaining database resources available, which can be grouped under the description *miscellaneous databases*. Note that the journal *Nucleic Acids Research* devotes its first issue every year to articles describing new databases and updates to existing ones. These articles can be accessed online at the following URL http:// www.nar. oupjournals.org/

Specialized Sequence Databases

The primary sequence databases are unbiased as to the type of sequence data they contain. However, a number of more specialized databases have been developed with particular types of nucleic acid or protein sequence in mind. For example, there are databases specifically for rRNA (ribosomal RNA) and tRNA (transfer RNA) sequences, for example the database of 5S rRNA sequences (http://biobases.ibch.poznan.pl/5SData/)and the database of small subunit rRNA sequences (http:/ /rrna.uia.ac.be/ ssu/). Further examples include databases for promoter sequences and other transcriptional regulatory elements, databases for regulatory elements in the noncoding region of mRNAs (messenger RNAs) and InBase, a database of inteins which are small peptides that are spliced out of some microbial proteins (http://www.neb.com/neb/inteins.html).

OMIM

OMIM (*Online Mendelian Inheritance in Man*) is a comprehensive database of human genes and genetic disorders maintained by the National Center for Biotechnology Information (NCBI) and can be accessed at the following URL http://www.ncbi. nlm.nih.gov/omim or through Entrez. Each OMIM entry has a full text summary of a gene or genetically determined phenotype and has numerous links to other databases such as the primary sequence databases, SWISS-PROT, PubMed references, general and locus-specific mutation databases, gene nomenclature databases and mapviewer. OMIM is an excellent starting point to find information on human genetics. An example of an OMIM file refers to the same gene/protein.

Incyte and UniGene

Incyte is an example of a *commercial database*. Unlike the *public databases* discussed above, which can be accessed freely by anyone using the WWW, commercial databases require subscription, as they are often the result of a single company's research and investment. *Incyte* is an integrated database of DNA sequences, transcripts, extensive annotations, expression data and access to cDNA (copy DNA) clones for experimental studies. It is the property of the LifeSeq® Foundation and subscription information can be found at the following URL:http://www.incyte.com.

UniGene is another resource for genome research. In the words of its developers: 'UniGene is an experimental system for automatically partitioning GenBank sequences into a non-redundant set of gene-oriented clusters. Each UniGene cluster contains sequences that represent a unique gene, as well as related information such as the tissue types in which the gene has been expressed and its map location'. UniGene incorporates about 10^5 expressed sequence tags (ESTs) and is used by experimenters to design probes and reagents for gene mapping and expression analysis. The organisms included in UniGene were chosen on the basis of the availability of large amounts of EST data and to give a reasonable coverage of the vertebrate and plant kingdoms with examples of closely and distantly related species. These include human, mouse, cow, rat, zebrafish, *Xenopus*, wheat, rice, barley, maize and *Arabidopsis*.

Structural Databases

Structural databases store data on protein (and nucleic acid) structure. The primary resource for protein structure data is the *Protein Data Bank* (*PDB*) available at the following URL: http://www.pdb.org/. This is the single worldwide archive of structural data and is maintained by the *Research Collaboratory for Structural Bioinformatics* (*RCSB*), at Rutgers University. The associated *Nucleic Acid Data Bank* (*NDB*) is also maintained there. Data from both X-ray crystallography and nuclear magnetic resonance (NMR) spectroscopy studies can be deposited in the PDB using a web-based interface called the *AutoDep Input Tool* (*ADIT*). The data are extensively checked and verified by human curators before acceptance. An equivalent European database is the *Macromolecular Structure Database* (*MSD*) maintained by the European Bioinformatics Institute. The RSCB and MSD databases contain the same data.

There are also other structural databases such as Entrez's *Molecular Modeling Database* (*MMDB*) which aims to provide information on sequence and structure neighbours, links between the scientific literature and 3D structures, and sequence and structure visualization.

Proteins and Higher-order Functions

One important aim of bioinformatics is to use biological data to understand the higher-level functions of the cell, that is, biochemical pathways, regulatory networks, signal transduction pathways, and how these influence cell and organism behaviour. A number of databases have been established with this goal in mind. Several databases have been designed, for example, to provide information on the functional annotation of proteins. Such databases include *PIR* (*Protein Information Resource*), which can be accessed at http://pir.georgetown.edu/. Another valuable resource is the *Kyoto Encyclopedia of Genes and Genomes* (*KEGG*), which is the primary resource of the Japanese GenomeNet service. KEGG is available at the following URL:http://www. genome.ad.jp/ kegg/. The main database integrates a number of subsidiaries including PATHWAY (which stores data on molecular pathways and complexes), GENES (which stores functional information about genes and their products) and LIGAND (which stores data about chemical compounds and reactions occurring in the cell). Together, these data can be used for functional annotation and the grouping of genes and proteins into common pathways, networks and hierarchies.

There are also many databases that provide information on specific aspects of protein function. For example, *DIP* (the *Database of Interacting Proteins*) and *BIND* (*Biomolecular Interaction Network Database*) provide functional annotations of proteins on the basis of their interactions with each other and with other ligands. There are also databases for particular types of protein, for example ReBase, a database of restriction endonucleases and their target sites; TRANSFAC, a database of transcription factors; Sentra, a database of signal transduction proteins,; and NUREBASE, a database of nuclear receptors.

Literature Databases

A literature database contains the abstracts and, in some cases, the full text and figures of published scientific articles. Such databases can be searched using text strings to find words in the title, abstract, keywords or main text, or by author or author's institution. One of the earliest comprehensive online library resources was *Medline*, which has been incorporated into a large resource called *PubMed* maintained by the NCBI. They are integrated into the Entrez suite of databases. Other such resources include the *Web of Science*, which requires institutional subscription, arrd *BioMedNet* (http://www.bmn.com), which provides access to thousands of review articles published in the popular *Trends* and *Current Opinion* journals. One of the best features of BioMedNet is that reviews can be downloaded as .pdf files and stored on the computer like a personal library.

Retrieval of Biological Data

Access to Distributed Data

A large amount of biological information is available over the World Wide Web, but the data are widely distributed and it is therefore necessary for scientists to have efficient mechanisms for *data retrieval*. One approach is to use standard *search engines* to find relevant web pages. However, it is sometimes difficult to find the desired information using this method, especially if the chosen search term has other connotations and pulls out many irrelevant sites. Alternatively, there are a number of dedicated *data retrieval tools* that can be used to access information for molecular biologists. The most widely used of these are *Entrez* and *DBGET* and SRS (sequence retrieval system). Each of these tools allows text-based searching of a number of linked databases as well as sequence searching with utilities such as BLAST. They differ in the databases they cover and how the retrieved information is accessed and presented.

Entrez

Entrez is a WWW-based data retrieval system, developed by the National Center for Biotechnology Information (NCBD, which integrates information held in all NCBI databases. These databases include nucleotide sequences (from GenBank and its subsidiaries), protein sequences, macromolecular structures and whole genomes. Other resources linked to the NCBI can also be searched using Entrez. These include OnLine Mendelian Inheritance in Man (OMIM) and the literature database MEDLINE, through PubMed. Entrez can be accessed via the NCBI web site at the following URL: http://www.ncbi.nlm.nih.gov/Entrez/. In total, Entrez links to 11 databases which are listed in Table 3.4.

Getting Started with NCBI and Entrez

Entrez is the common front-end to all the databases maintained by the NCBI and is an extremely easy system to use. The Entrez main page, as with all NCBI pages, is undemanding

Table 3.4. The databases covered by Entrez, listed by category

Category	*Database*
Nucleic acid sequences	Entrez nucleotides: sequences obtained from GenBank, RefSeq and PDB
Protein sequences	Entrez protein: sequences obtained from SWISS-PROT, PIR, PRF, PDB, and translations from annotated coding regions in GenBank and RefSeq
3D structures	Entrez Molecular Modelling Database (MMDB)
Genomes	Complete genome assemblies from many sources
PopSet	From GenBank, set of DNA sequences that have been collected to analyse the evolutionary relatedness of a population
OMIM	Online Mendelian Inheritance in Man
Taxonomy	NCBI Taxonomy Database
Books	Bookshelf
ProbeSet	Gene Expression Omnibus (GEO)
3D domains	Domains from the Entrez Molecular Modelling Database (MMDB)
Literature	PubMED

in its browser requirements and downloads quickly. The databases available for searching can be accessed by hyperlinks at the top of the page, or by using the drop-down menu as shown. Once a database has been selected, a *search term* is then entered in the space provided. The search term may be a single word or a Boolean phrase. Clicking on 'GO' initiates the search. Hits in the selected database are displayed (these are known as *neighbours*) and matching records in other Entrez databases are also shown (these are known as *links*). Hits are ordered by similarity based on *precomputed analysis* of sequences/structures or the literature.

For the newcomer, the following URL provides an overview of Entrez and a useful tutorial: http://www.ncbi.nlm.nih.gov/Database/index.htrnI. This page and includes a diagram showing the connectivity between eight of Entrez's databases.

DBGET/LinkDB

DBGET is an integrated data retrieval system developed and jointly maintained by the Institute for Chemical Research (Kyoto University) and the Human Genome Center (University of Tokyo).

It is integrated with more than 20 data bases, which can be searched one at a time or in combination using the commands *bfind* (for text searches) or *bget* (for searches based on accession number). Hits are presented as a list of results together with any available associated information.

Table 3.5. The databases covered by DBGET/LinkDB, listed by category

Category	*Database*
Nucleic acid sequences	GenBank, EMBL
Protein sequences	SWISS-PROT, PIR, PRF, PDBSTR
3D structures	PDB
Sequence motifs	PROSITE, EPD, TRANSFAC
Enzyme reactions	LIGAND
Metabolic pathways	PATHWAY
Amino acid mutations	PMD
Amino acid indices	AAindex
Genetic diseases	OMIM
Literature	LITDB Medline
Organism-specific gene catalogs	*E. coli, H. influenzae,* M. *geni-talium,* M. *pneumoniae* M. *jann-aschii, Synechocystis,* S. *cerevisiae*

LinkDB is an associated database of links (*binary relation-ships*) between entries in the different databases available to DBGET and further organism-specific databases, such as the C. *elegans* Database (ACeDB), Flybase and the *Saccharomyces* Genome Database (SGD). DBGET is closely associated with KEGG, the Kyoto Encyclopedia of Genes and Genomes, which is maintained by the same group.

Data Retrieval with SRS

SRS and Entrez/DBGET

SRS (*sequence retrieval system*) is a retrieval tool developed by the European Bioinformatics Institute (EBI) that integrates over 80 molecular biology databases. These are listed at http:// srs6.ebi.ac.uk/srs6bin/cgi-bin/wgetz?-page+ databanks+-newId, and are summarized in Table 3.6. Like Entrez and DBGET, SRS can be used over the WWW. However, the difference between SRS and Entrez/DBGET is that SRS is open source software that can be downloaded and installed locally. The result is that the

Table 3.6. The databases covered by the SRS at http://srs6.ebi.ac.uk, listed by category

SRS description	*Examples*
Literature	MEDLINE, GO, GOA
Sequence	EMBL, EMBLNEW, SWISSPROT, SPTREMBL, REMTREMBL, TREMBLNEW, ENSEMBL, PATENT_PRT, USPO_PRT, IMGTLIGM, IMGTHLA
InterPro&Related	INTERPRO, IPRMATCHES, IPRMATCHES _ENSEMBL, PROSITE, PROSITEDOC, BLOCKS, PRINTS, PFAMA, PFAMB, PFAMHMM, PFAMSEED, PRODOM
SeqRelated	UTR, UTRSITE, TAXONOMY, GENETICCODE, EPD, HTG_QSCORE, CPGISLAND, EMBLALIGN, EMESTLIB
TransFac	TFSITE, TFFACTOR, TFCELL, TFCLASS, TFMATRIX, TFGENE
User Owned Databanks	USERDNA, USERPROTEIN
Application Results	FASTA, FASTX, FASTY, NFASTA, BLASTP, BLASTN, CLUSTALW, NCLUSTALW, PPSEARCH, RESTRI-CTION MAP
Protein3DStruct	PDB, DSSP, HSSP, FSSP, PDBFINDER, RESID
Genome	MOUSE2HUMAN, LOCUSLINK, HGNC, HSAGENES
Mapping	RHDB, RHDBNEW, RHEXP, RHMAP, RHPANEL, OMIMMAP
Mutations	OMIMALLELE, MUTRES, SWISSCHANGE, EMBLCHANGE, MUTRESSTATUS, OMIM, OMIMOFFSET, HUMUT, HUMAN_MITBASE, P53L1NK
Locus Specific Mutations	41 *entries omitted here*
SNP	MITSNP, dbSNPSubmitter, dbSNPAssay, dbSNPSNP, HGBASE, HGBASE_SUBMITTER
Metabolic Pathways	LENZYME, LCOMPOUND, PATHWAY, ENZYME, EMP, MPW, UPATHWAY, UREACTION, UENZYME, UCOMPOUND, UIMAGEMAP
Others	REBASE, SRSFAQ, BIOCATAL
System	PRISMASTATUS

databases and utilities that are available to the end user are not restricted by the activities of curators [at the National Center for Biotechnology Information (NCBI) in the case of Entrez, or GenomeNet in the case of DBGET]. Several large sites, including SWISS-PROT, use SRS as standard. The main ExPASy site (www.expasy.ch). one of the most useful bioinformatics gateways is an example of SRS in action.

Using SRS

To start using SRS, access the appropriate site on your local network (if SRS has been installed locally) or alternatively try the SRS homepage at http://srs6.ebi.ac.uk. On the WWW version, the top page lists 17 *classifications*. By clicking on the adjacent '+' symbol, each classification can be expanded to reveal the associated databases, and by clicking on the '–' symbol, the classifications can be collapsed. It is also possible to expand collapse all classifications. Note that SRS classifications, unlike those in Entrez, are grouped by the type of data not the type of molecule. Therefore, the *Sequence Libraries* classification covers all sequences (nucleotide and protein) whereas these data are separated by Entrez. Most of the other classifications are self-explanatory, but new users may find some unfamiliar. For example, *InterPro&Related* refers to the secondary databases of protein motifs, *SeqRelated* refers to specialized sequences databases such as UniGene and *TransFac* refers to transcription factor databases.

To search with SRS, expand the relevant classifications and check boxes corresponding to the required databases. Clicking on the 'standard' or 'expanded' buttons then brings up the query form ('standard' is recommended for newcomers). Search terms can be entered in a number of fields by selecting from the drop-down menu (e.g. Accession number Description, Keywords, Organism) or alternatively it is possible to search all fields simultaneously. Up to four different search terms can be used, linked by Boolean operators across multiple fields, so quite specific searches can be carried out. After clicking the 'submit query' button, the query results are displayed as a table, with hyperlinks to files in the appropriate databases. A helpful tutorial for SRS users is available at http://srs6.ebi.ac.uk. From here, follow the "Information" link and look at the "User Guide".

Installing SRS

SRS is available to be installed locally. Essentially, it provides both the graphical interface and files to allow databases to be accessed. SRS works by constructing indexing files and should be updated regularly. The installation set comes with several configuration files and module files for the most popular bioinformatic databases ('databanks' in SRS). SRS configuration files establish which databases are available to SRS. Some of these files are used to add databases and others specify a grouping mechanism for databases or create new built-in views for data results. Module files correspond to the databases to be indexed.

Thus, an SRS system can be installed and operated with the minimum of effort. However, there are several points to consider when planning (or asking for help with) an SRS installation. Although the databases may all be remote, the indexing files are very large and, on a PC or other small system, the indexing process can take a lot of processor time (days). Nevertheless, a modest SRS system using example configuration module files is easily maintained and updated. The strength of SRS lies, in part, with providing the installer the capacity to use databases of his or her own even if they use a novel format. In order to do this, SRS has an associated scripting language called *Icarus*.

4

SEARCHING DATABASES

SEQUENCE SIMILARITY SEARCHES

The sequences of biological macromolecules in modern databases are the products of molecular evolution. When sequences share a common ancestral sequence, they tend to exhibit similarity in their sequences, structures and biological functions. This is probably the most powerful idea in bioinformatics, because it enables us to make predictions. Often little is known about the function of a new sequence from a genome sequencing program, but if similar sequences can be found in the databases for which functional or structural information is available, then this can be used as the basis of a prediction of function or structure for the new sequence. It is very useful, therefore, to have search tools that take the new sequence as 'query', and search the database for similar sequences. Examples of such tools are the BLAST, FASTA and Smith-Waterman algorithms, which we will discuss later, but first we need to discuss how sequence similarity might be discovered, visualized and quantified.

Sequence Alignment

Any pair of nucleic acid sequences will share a degree of similarity. For instance, DNA sequences are constructed from an alphabet of only four letters (A, T, G, C), so any sequence that consists of a mixture of these letters will show some similarity to any other similarly constructed sequence. We need to distinguish between this type of 'chance' similarity, and the similarity that is the result of a real evolutionary and/or

SEQ1: AATTGATTGCGCATTTAAAGGG
SEQ2: AACTGA---CGCATCTTAAGGG

Fig. 4.1. An alignment of two short DNA sequences.

functional relationship. We need a way of quantifying similarity. Quantifying similarity begins with an *alignment*, such as that shown for the very short DNA sequences in Fig. 4.1. In this figure, the two sequences are written one above the other and the letters (defining DNA bases), which are vertically directly above and below each other are *aligned* or *equivalenced*. Reading the alignment from left to right, the first two As in each sequence are aligned, and these are followed by a T aligned with a C and then another pair of aligned T_s and so on. It is important that not all letters in a particular sequence have equivalents in the other sequence. In Fig. 4.1, letters 7-9 of sequence 1 (TTG) are not aligned with any letters from sequence 2. When this happens, we say that a *gap* has been introduced.

The point of introducing a gap here is that it enables a better alignment of the two sequences, in which more of the aligned pairs are identical letters. In this case, it enables the shared CGCAT directly after the gap to align between the two sequences. Alignments can be interpreted in evolutionary terms. When identical letters are aligned, the simplest interpretation is that these letters were part of the ancestral sequence and have remained unchanged. When non-identical letters are aligned, the simplest interpretation is that a mutation has occurred in one of the sequences.

Of course, in the absence of information about the ancestral sequence, it is not possible to know in which sequence the mutation actually occurred. Gaps in alignments can be interpreted in terms of the *insertion* or *deletion* of letters in one of the sequences with respect to the ancestral sequence. In Fig. 4.1, the gap could have resulted either from an insertion of three letters in sequence 1 or a deletion of three in sequence 2. Again, without the ancestral sequence, these possibilities cannot be distinguished, so the gap is sometimes referred to as an *indel*.

Alignment Algorithms

Finding the best alignment between the two sequences in Fig. 4.1 was fairly straightforward, because the sequences are short and very closely related. It is much more difficult if the sequences are longer or if they are less closely related. In these cases, computational methods are required to find the best alignment of the sequences. Fortunately, there are known computational methods for this task, called *dynamic programming algorithms*. These algorithms take two input sequences and produce as output the best alignment between them.

Sequence similarity analyses commonly use two dynamic programming algorithms, the *Needleman-Wunsch algorithm* and the *Smith-Waterman algorithm*. These are closely related, but the main difference is that the Needleman-Wunsch algorithm finds *global* similarity between sequences, while the Smith-Waterman algorithm finds *local* similarity. An alignment from the Smith-Waterman algorithm might only cover a small (local) part of each sequence, while a Needleman-Wunsch alignment will try to cover as much of the sequences as possible, starting at left-most end of one of the sequences and finishing at the right-most end of one of the sequences.

The Smith-Waterman algorithm is the most used because real biological sequences are often not similar over their entire lengths, but only in local portions. Examples of biological reasons for this are genes from different organisms with similar exons (local similarity) but different intron structures, or proteins that share only certain domains. Even though the Smith-Waterman algorithm is able to find these local similarities, it should also discover global sequence similarity if it exists.

Alignment Scores and Gap Penalties

In their simplest incarnation, dynamic programming algorithms find alignments containing the largest possible number of identically matching letters, as in the alignment of Fig. 4.1. The problem, however, is that if it is permitted to insert gaps freely into the alignment then the algorithms produce alignments containing very large proportions of matching letters and large numbers of gaps. We need to control this, because insertion and deletion of monomers is a relatively slow evolutionary process, and alignments with large numbers of gaps do not make biological sense. Dynamic programming algorithms get around this problem by using *gap penalties*.

Essentially the algorithms form a score reflecting the quality of the alignment (a high positive score indicates a good alignment). A simple score contains a positive additive contribution of 1 for every matching pair of letters in the alignment, and a gap penalty is subtracted for each gap that has been introduced. In Fig. 4.1, if the gap penalty were equal to 1, then because there are 16 identically matching letters in the alignment and one gap, the score would be 16 -1 = 15.

Dynamic programming algorithms find the alignment between the two sequences that maximizes the alignment score. This alignment depends on the choice of gap penalties: high gap penalties result in shorter, lower-scoring alignments with fewer gaps and lower penalties give higher-scoring, longer alignments with more gaps. Several forms of gap penalty are commonly used. We described the simplest form above, where each gap attracted a *constant penalty* of I, independent of the length of the gap. In general, a constant gap penalty can be written as A (A= 1 above) where the size, of A controls how strongly gaps are penalized. Sometimes a *proportional penalty* is used, where the penalty is proportional to the length of the gap. This penalty has the form Bl, where B is a constant (again controlling how strongly gaps are penalized), and l is the length of the gap. With this form, longer gaps attract larger penalties than shorter ones, which makes good biological sense.

Finally, the most complex form of gap penalty is known as affine; this has both constant and proportional contributions, and it takes the form $A+Bl$. In this case, the constant A is called the *gap-opening penalty*, because it is applied to a gap of any length. The constant B is called the *gap-extension penalty*, because it is the penalty attached to extending the length of an existing gap by one unit. The motivation for the affine gap penalty is that opening a gap (penalty=A) should be strongly penalized, but once a gap is open it should cost less (B) to extend it. It is not always clear what values should be used for the gap penalty constants A and B, and it does depend to some extent on the intended application.

Most software for sequence alignment or similarity search comes with good default values for general sequence alignment or searching. In general, if you are interested in detecting only close sequence relationships then it would be a good idea to try increasing gap penalties above the default values to reduce the number of gaps. On the other hand, more distant relationships might be discovered by decreasing gap penalty values. It is often worth experimenting with different gap penalties around the default values. In some applications, gap penalties that differ significantly from default values should be used. One example would be the removal of vector sequence during genome sequencing: here the concern is only to find exact matches to the known Vector sequence and very high gap penalties are appropriate.

Measuring Sequence Similarity

We have explained how dynamic programming algorithms attribute an alignment score that reflects the degree of relatedness of the aligned sequences. Some other measures are used, and perhaps the most common is the *percentage of identically aligned residues*. In Fig. 4.1, the number of identically aligned residues is 16, and the length of the alignment is 22, so the percentage sequence identity is (16/22) x 100 = 73%. This has the advantage that it is independent of the length of the alignment, so it can provide comparability between alignments of different length. However, it must always be remembered that high percentage identities are much more likely to reflect real biological or evolutionary relationships if they extend over long alignments.

An alignment with 50% identity means little if its length is only 10 nucleotides, because it is likely to have occurred by chance. An alignment of 50% identity over 100 nucleotides is much more likely to be significant. Perhaps the most powerful methods of assessing sequence similarity are the statistical measures. Aligning sequences to maximize identical matches is reasonable for DNA sequences. However, with protein sequences, it is useful to take more account of the properties of the monomers that make up the sequence. Some amino acids substitute for each other better than others, and it is useful to take account of this in protein sequence alignment algorithms.

Similarity and Homology

These terms are often confused. Any set of sequences can exhibit similarity and this may be quantified as discussed above. Sequences are *homologous* only if they evolved by divergence from a common ancestor. To say that two sequences share 50% homology is nonsense; to say that they share 50% similarity and that this indicates possible homology is the correct usage of the terms. Sequence similarity searches are used very commonly to predict gene or protein function. The underlying theory is that similar sequences are likely to be homologous and therefore to have similar functions. If a biological function is essential for an organism to survive and reproduce, then loss of that function would be selected against by evolution, and this is the reason why homologous genes might be expected to retain the same function in different organisms. However, the evolutionary mechanism of gene duplication allows organisms to acquire redundant copies of genes. These redundant copies are then free to evolve new functions, and become homologous genes with different functions.

This is the principle objection to prediction of gene function through sequence similarity, and is probably the reason for a significant amount of incorrect bioinformatics-derived annotation in the sequence databases. Examples of homologous genes with different functions are lysozyme (an enzyme) and α-lactalbumin (a mammalian regulatory protein). These proteins have very similar sequences, are almost certainly homologous and yet have very different functions. When two homologous genes in different species have the same function, they are known as *orthologs;* when two genes in the same or different species have different functions they are known as *paralogs*.

AMINO ACID SUBSTITUTION MATRICES

Maximizing Amino Acid Identities

Protein sequences are constructed from an alphabet of 20 naturally occuring amino acids. Like nucleic acid sequences, protein sequences can be aligned to maximize the number of identically matching pairs within the alignment. This is a good way of aligning closely related sequences. In this case, the contribution of every aligned pair of identical amino acids to the alignment score is one, and the contribution of every aligned pair of non-identical amino acids is zero.

Evolution

Evolution changes nucleic acid sequences, but protein-coding sequences evolve under strong functional constraints. Changes to these sequences generally survive only if they do not have a deleterious effect on the structure and function of the protein, because loss of the function of a protein is usually a disadvantage to the organism. An obvious exception to this occurs after a gene duplication event, when one copy of the gene can evolve very quickly, perhaps to become a pseudo-gene, or perhaps to gain a new and useful function. Usually, however, protein-coding genes

C	1																			
S	0	1																		
T	0	0	1																	
P	0	0	0	1																
A	0	0	0	0	1															
G	0	0	0	0	0	1														
N	0	0	0	0	0	0	1													
D	0	0	0	0	0	0	0	1												
E	0	0	0	0	0	0	0	0	1											
Q	0	0	0	0	0	0	0	0	0	1										
H	0	0	0	0	0	0	0	0	0	0	1									
R	0	0	0	0	0	0	0	0	0	0	0	1								
K	0	0	0	0	0	0	0	0	0	0	0	0	1							
M	0	0	0	0	0	0	0	0	0	0	0	0	0	1						
I	0	0	0	0	0	0	0	0	0	0	0	0	0	0	1					
L	0	0	0	0	0	0	0	0	0	0	0	0	0	0	0	1				
V	0	0	0	0	0	0	0	0	0	0	0	0	0	0	0	0	1			
F	0	0	0	0	0	0	0	0	0	0	0	0	0	0	0	0	0	1		
Y	0	0	0	0	0	0	0	0	0	0	0	0	0	0	0	0	0	0	1	
W	0	0	0	0	0	0	0	0	0	0	0	0	0	0	0	0	0	0	0	1
	C	S	T	P	A	G	N	D	E	Q	H	R	K	M	I	L	V	F	Y	W

Sequence 1	:	MILVKP	VVLKGDFG
Sequence 2	:	MILLKP	AIIIRAEY
Position score	:	111011	00000000

Total alignment score = sum of position scores =5.

Fig. 4.2. Alignment to maximize amino acid identities, and the associated substitution score matrix.

evolve much more slowly than most other parts of any genome, because of the need to maintain protein structure and function.

Allowed Changes

When evolutionary changes do occur in protein sequences, they tend to involve substitutions between amino acids with similar properties, because such changes are less likely to affect the structure and function of the protein. For instance, hydrophobic amino acids of similar size tend to substitute for each other quite well, because, more often than not, these occupy positions within the hydrophobic core of the protein, where tight packing and hydrophobicity strongly affect the stability of the protein structure. Examples of such mutations would be changes between LEU and ILE or VAL. Protein sequences from within the same evolutionary family usually show substitutions between amino acids with similar physicochemical properties.

Substitution Score Matrices

Substitution score matrices are used to show scores for amino acid substitutions. For example, Fig. 4.2 shows a very simple score matrix, in which identical substitutions score 1 and nonidentical

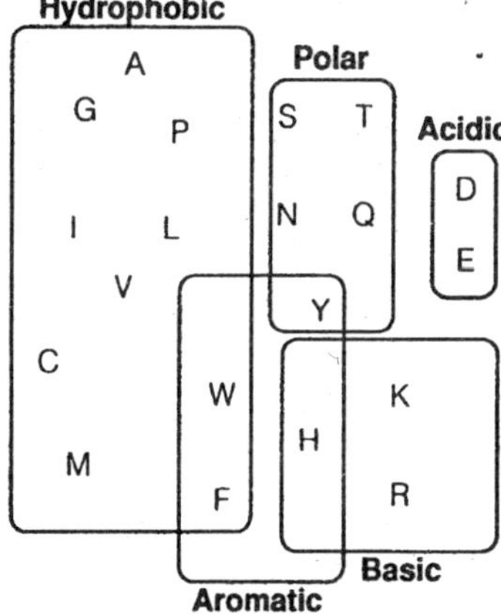

Fig. 4.3. Relationships between the physicochemical properties of amino acids.

ones score zero. More general matrices show the likelihood of substitution occurring between each pair of amino acids. They enable the detection of similarity between protein sequences that would be missed by the simple identity matrix of Fig. 4.2. An example is the *PAM250 substitution matrix*: its elements are scores for alignment or (i.e. substitution between) each possible pair of amino acids. For instance, the score for aligning the similar amino acids L and I is high (2), reflecting the high likelihood of this substitution as an evolutionary process while the score for aligning the dissimilar amino acids G and W is low (–7). Notice that the scores for alignments of identical amino acids vary. These reflect the frequency of occurrence of the amino acids in natural protein sequences.

The alignment of two identical uncommon amino acids (e.g. aligning W with W) is more likely to reflect an evolutionarily significant alignment than the alignment of two identical common ones (e.g.S with S), which would be likely to happen by chance in two unrelated protein sequences. Alignments of uncommon identical amino acids are therefore given higher scores. The dynamic programming sequence alignment algorithms are not limited to aligning identical letters. They can operate with any system of scoring alignments between monomers that can be expressed as a substitution matrix. The only difference is that instead of scoring 1 for every match and 0 for every mismatch, substitution scores are taken from the matrix. The total alignment score that the algorithms maximize is the sum of the

	C	S	T	P	A	G	N	D	E	Q	H	R	K	M	I	L	V	F	Y	W
C	12																			
S	0	2																		
T	-2	1	3																	
P	-1	1	0	6																
A	-2	1	1	1	2															
G	-3	1	0	-1	1	5														
N	-4	1	0	-1	0	0	2													
D	-5	0	0	-1	0	1	2	4												
E	-5	0	0	-1	0	0	1	3	4											
Q	-5	-1	-1	0	0	-1	1	2	2	4										
H	-3	-1	-1	0	-1	-2	2	1	4	3	6									
R	-4	0	-1	0	-2	-3	0	-1	-1	1	2	6								
K	-5	0	0	-1	-1	-2	1	0	0	1	0	3	5							
M	-5	-2	-1	-2	-1	-3	-2	-3	-2	-1	-2	0	0	6						
I	-3	-1	0	-2	-1	-3	-2	-2	-2	-2	-2	-2	-2	2	5					
L	-6	-3	-2	-3	-2	-4	-3	-4	-3	-2	-2	-3	-3	4	2	6				
V	-2	-2	0	-1	0	-1	-2	-2	-2	-2	-2	-2	-2	2	4	2	4			
F	-4	-3	-3	-5	-4	-5	-4	-6	-5	-5	-2	-4	-5	0	1	2	-1	9		
Y	0	-3	-3	-5	-3	-5	-2	-4	-4	-4	0	-4	-4	-2	-1	-1	-2	7	10	
W	-8	-2	-5	-6	-6	-7	4	7	7	5	3	2	-3	-4	-5	-2	-6	0	0	17

Sequence 1 : MILVKP –VVLKGDFG
Sequence 2 : MILLKP A I I I RAEY–
Position score : 656556 0 4 4 2 3 1 3 7 0
Total alignment score = sum of position scores) – (gap penalty) = 54 – 1 = 53.

Fig. 4.4. The PAM250 matrix and the alignment of the sequences from fig 4.2 produced with this matrix.

substitution scores in the alignment, with the usual gap penalty contribution. PAM stands for 'accepted point mutations'; the P and A are swapped to make it easier to say. PAM250 refers to an evolutionary distance of 250 accepted point mutations (PAMs) per 100 amino acid residues. PAM matrices are derived by counting observed evolutionary changes in closely related protein sequences, and then exuapolating the observed transition probabilities to longer evolutionary distances.

It is possible to derive PAM matrices for any evolutionary distance, but in practice, the most commonly used matrices are PAM120 and PAM250. PAM matrices with smaller numbers represent shorter evolutionary distances. Choosing the matrix for the most appropriate evolutionary distance might result in the best possible alignment of two sequences, but in practice, it is rarely possible to know what the evolutionary distance is, and experience shows that the PAM250 matrix usually produces reasonable alignments.

The *BLOSUM* series of matrices, and in particular BLOSUM50 and BLOSUM62, have recently become popular. These matrices have some theoretical advantages (their derivation does not involve the process of extrapolation to large evolutionary distances), and there is some evidence to suggest that they do outperform PAM matrices in. practice. For this reason they are now more commonly used. The numbers attached to BLOSUM matrices do not have the same interpretation as those for PAM matrices (accepted mutations per 100 residues). In fact, BLOSUM matrices with smaller numbers represent longer evolutionary distances (BLOSUM50 represents a longer distance than BLOSUM62). There are several other matrices in fairly common use, and some specialized matrices (for instance PAM matrices specific to integral transmembrane proteins have been derivied).

Significance

A great deal of bioinformatics is concerned with the detection of evolutionary relationships between sequences. The use of matrices like BLOSUM62 extends our ability to detect distant relationships far beyond what could be found using the identity matrix. This ability to detect very distant relationships (sometimes between sequences whose percentage of identical residues has diverged to below 30%) is the reason why sequence comparison should always be carried out at the protein rather than the nucleic acid sequence level, when this is possible. The ability to encode permissible changes in protein structures means that protein sequence alignment can reveal much more distant evolutionary relationships than naive comparison of nucleic acid sequences.

Visualization

It is often useful to be able to visualize the similarity between two sequences. This can be done using dot plots. In the case, the dot plot has been used to discover internal similarity between N- and C-terminal portions of the same sequence, but it is equally possible to use dot plots to study similarity between two different sequences.

Database Searches: FASTA And Blast

Database Searches

Sequence similarity searches aim to extract sequences from a database that are similar to a query sequence, perhaps with the aim of gathering family members, or predicting structure and function. As already discussed ways of quantifying the similarity

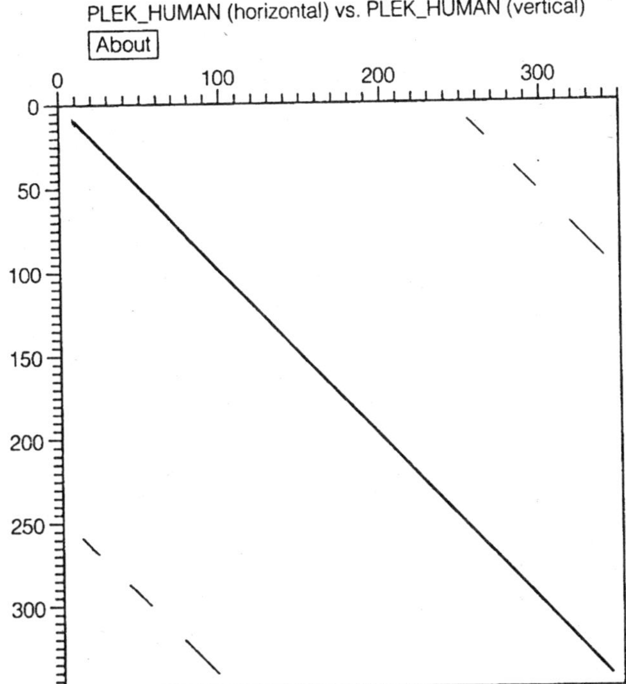

Fig. 4.5. A dot plot of the human pleckstrin sequence against itself produced with Erik Sonnhammer's 'dotter' program.

between macromolecular sequences using sequence alignments and substitution score matrices. Similarity searches work by comparing (aligning) the query sequence to each database sequence in turn and then ranking the database sequences with the highest scoring (most similar) at the top. This process can be carried out by the dynamic programming algorithms, but in practice, these algorithms are often too slow for searching large databases. We will discuss some alternative methods in this section. Also discussed in this section are the statistics of sequence similarity searching, and some issues surrounding database types and search parameters.

Algorithms and Software

Dynamic programming algorithms are guaranteed to find the best alignment of two sequences for given substitution matrices and gap penalties. This is impressive, but the process is often quite slow, perhaps taking hours for a search of a large database.

For this reason, alternative methods have been developed. Perhaps the most used of these are *FASTA* and *BLAST*; they are typically five (FAST A) to 50 (BLAST) times faster than dynamic programming. Unlike dynamic programming, these methods are not guaranteed to find the best alignment, and for this reason they could be less accurate. In practice, however, BLAST and FAST A usually suffer a minor degradation in accuracy when compared with dynamic programming, and the increased speed is often worth this small cost.

FAST A and BLAST are both based around similar ideas; in particular, they both make the reasonable assumption that high-scoring alignments are likely to contain short stretches of identical or near identical letters. These short stretches are called words. In the case of BLAST, the first step is to look for words of a certain fixed word length W that score higher than a certain threshold score (*T*). The value of *W* is normally 3 for protein sequences or 11 for nucleic acid sequences. *W* and *T* are under the control of the user, but there is rarely any need to change the default values. FAST A, on the other hand, looks for identically matching words of length *ktup*. *ktup* is under the control of the user, but the default values of 2 for protein sequences and 6 for DNA sequences are again seldom changed.

Both FAST A and BLAST employ a second step where extensions to the initially identified word matches are sought. BLAST extends individual word matches until the total score of the alignment falls from its maximum value by a certain amount producing an alignment without gaps. Early versions of BLAST always produced alignments without gaps, but later versions now produce gapped alignments. The BLAST terminology for these high-scoring local alignments is *high-scoring segment pairs* (*HSPs*).

FAST A tries to find ungapped alignments of the sequence that contain a high density of the initially identified word matches, and then attempts to join these into high-scoring gapped alignments. Finally, having identified high-scoring alignments between the sequences, FASTA produces its final alignment and score by a full dynamic programming alignment of the identified high-scoring regions of the sequences. While BLAST might identify several regions of high-scoring similarity between two sequences, FASTA only identifies the single highest scoring region.

We have already commented that the principal advantage of FASTA and BLAST over dynamic programming is that of speed. This has been very important for those developing WWW access to sequence database searches. When a database is accessed regularly by users from all over the world, it is not uncommon that the computers running the searches have several jobs running at the same time. When this happens even the most powerful computers can have problems, and the amount of computational resource required by each search takes on crucial importance. For this reason, WWW developers have used the BLAST and FASTA algorithms in preference to full dynamic programming with the Smith-Waterman algorithm.

Statistical Scores

One of the main problems with sequence similarity searches is to know when an identified sequence similarity is significant, that is, when it represents a real biological or evolutionary relationship rather than a chance similarity between unrelated sequences. We already discussed how alignments of protein sequences could be scored using a combination of substitution matrices and gap penalties. The higher the score of an alignment by this method the more likely it is to reflect a real biological relationship, but how high is a high score?

A major breakthrough in sequence similarity searches was the development of a statistical theory of the alignment scores. Like many theories of statistical significance, this theory centers on the calculation of *p values*. In the sense of a sequence similarity search, the p value of an identified similarity of score S is the probability that a score of at least S (i.e. a score of S or greater) would have been obtained in a chance match between two unrelated sequences of similar composition and length. Very low values of p therefore correspond to significant matches: in these cases it is highly improbable that the score obtained occurred by chance, and it is probable that it occurred as the result of a real biological or evolutionary relationship. All the search methods we have discussed so far report p values for identified similarities, and these can be used to judge the significance of the discovered relationships. Of course, the user still has to form a judgment about his/her own threshold p value for considering a similarity significant. Clearly a value of $p=O.5$ does not represent a significant similarity (because it has a 50% probability of occurring by chance between unrelated sequences), but values of p less than 0.01 are much more significant.

A related quantity that is reported by most search software is the E (*Expect*) *value*. For an identified similarity of score S, the E value is the expected frequency of scores of at least S. It can be interpreted as the number of scores of at least s that would have been expected to have occurred by chance. The calculation of p and E values can be carried out for the two matched sequences in isolation (i.e. the probability of a match between the two sequences scoring at least as high by chance) or for the entire database search (i.e. the probability of a score at least as high in the entire database search). The latter are generally higher (because the database is larger than any individual sequence), and are obviously the most relevant in assessing database search results. FASTA and BLAST take slightly different approaches to whole database p and E values.

The E values reported by the FAST A software are related to the p value by the equation $E=Np$, where N is the number of sequences in the database (i.e. E refers to the database search but p to the sequences in isolation). This reflects an assumption that all the database sequences are equally likely to be related to the query sequence, which is questionable when they may differ significantly in length. The calculation of E values for BLAST does not make this assumption, but rather treats the database as one very long sequence. Nevertheless, the FASTA calculation illustrates the following important fact: even when the p value for a match between two sequences is low, E values for search of a large database can be quite large. For instance a FASTA similarity score S with a p value of 0.001 might seem fairly significant, but in a database of 10000 sequences this would correspond to an E value of $0.001 \times 10000 = 10$. So, we would expect the scores of at least S to occur 10 times by chance, which is clearly not a high level of significance. For this reason E values are often more informative than p values.

Sensitivity and Specificity

Evaluation of the results of a database search is best viewed in terms of two complementary measures, the *sensitivity* and the *specificity*. Suppose a threshold value of E or p is chosen so that sequence similarities corresponding to p or E values lower than the threshold are considered significant. It is common to call the sequences with significant similarities *hits*. The search partitions the database into two subsets, hits (positives) and non-hits (negatives). These can be further partitioned, at least conceptually, into true and false positives and true and false negatives. A true positive is a hit with a real biological relationship to the query sequence, and a false positive is a hit without such a relationship. A true negative is a non-hit with no real biological relationship to the query, and a false negative is a non-hit with a real biological relationship to the query. The sensitivity of the search is the proportion of the real biological relationships in the database that were detected as hits, and it can be written as

$$S_n = n_{tp}/(n_{tp} + n_{fn})$$

where n_{tp} is the number of true positives and n_{fn} is the number of false negatives. The *specificity* of the search is the proportion of hits that correspond to real biological relationships, and it can be written as

$$S_p = n_{tp}/(n_{tp} + n_{fp})$$

where n_{fp} is the number of false positives. An ideal database search would have sensitivity and specificity both as close as possible to 1, but in practice, this is not attainable. Note that there is a trade-off between the two quantities. If the threshold for significance is increased, so that more hits are considered significant, then the sensitivity is likely to increase (there will be more true positives and less false negatives), but the specificity is likely to decrease (there will be more false positives).

It is of course only possible to carry out an analysis of sensitivity and specificity if the real biological relationships in the database are already known (so that they can be assigned to the categories of true and false positive and negative). Further, this information must come from a source independent of sequence similarity analyses, for instance experimental determination of protein structure and function. In general, this information is not available, but it is useful to understand the trade-off between sensitivity and specificity when choosing threshold p and E values. Some investigators have used ideas like these with sets of proteins of known structure and function to compare the performance of FAST A, BLAST and Smith-Waterman methods, and have found evidence for slightly better performance of the Smith-Waterman by comparison with the faster methods.

Database Types

The database search methods we have discussed can be used with protein or nucleic acid databases and query sequences. Several of these programs first translate a nucleic acid sequence in all possible reading frames to produce six possible coded protein sequences. Sequences are then compared at the protein level rather than the nucleic acid level. When these options are chosen in preference to options that compare simply at the nucleic acid level, the searches take longer to run but the results are better. This is because it is possible to detect much more distant relationships using protein sequences.

Table 4.1. BLAST and FASTA programs ('translated' means the nucleic acid sequence is translated to protein in six reading frames)

Program name	*Query sequence*	*Database type*
Blastp	Protein	Protein
Blastn	Nucleic acid	Nucleic acid
Blastx (translated)	Protein	Nucleic acid
Tblastn	Nucleic acid (translated)	protein
Tblastx (translated)	Nucleic acid (translated)	Nucleic acid
Fasta	Protein or nucleic acid	Protein or nucleic acid
Tfastx	Nucleic acid (translated)	protein
Fastx (translated)	Protein	Nucleic acid

Worked Example and Program Availability

If you want to use BLAST or FAST A to search the large, general-purpose sequence databases like SWISS-PROT, GenBank and that at the European Molecular Biology Laboratory (EMBL) then it is probably best done using the WWW sites of the organizations that maintain these databases, that is, the European Bioinformatics Institute (EBI) and the National Center for Biotechnology Information (NCBI). Appropriate links are available on the text WWW site. Here we give a worked example using FASTA at the EBI site and BLAST at the NCBI site.

The search carried out was with the sequence of human pleckstrin and similar sequences were sought in the SWISS-PROT database. The searches were carried out with the BLOSUM62 amino acid substitution matrix and gap-opening and gap-extension penalties of 8 and 2. These relatively low penalties were used to achieve maximum comparability between the two search servers using the limited sets of values available at each site, and ideally the higher default penalties of 12 and 2 should be used. Using 0.1 as a threshold *E* value below which similarity was considered significant, the algorithm found 17 significant sequence similarities. Note that the alignment covers almost the entire length of the pleckstrin sequence.

Note that in this case BLAST finds more similarities below the *E*=0.1 cutoff chosen. BLAST and FASTA use different ways of calculating *E* values and it is unsurprising that they produce different results. Both algorithms find the significant similarity to the human cytohesin 1 sequence featured. BLAST finds two regions of similarity, but unlike FASTA does not find similarity over the entire sequence.

In the case of these sequences, it is instructive to compare the results with what is already known about the sequences. It is known that the human pleckstrin sequence contains two pleckstrin homology (PH) domains (amino acids 5-100 and 245-345) separated by another unrelated domain. Cytohesin 1 contains a PH domain (amino acids 260-375) preceded by an unrelated domain. The first of the BLAST alignments corresponds to the match between second PH domain of pleckstrin and the PH domain of cytohesin, and this similarity is also the part of the FASTA alignment where similarity is strongest. The second BLAST alignment is the similarity between the first PH domain of pleckstrin and the PH domain of cytohesin. BLAST does not discover any sequence similarity outside the PH domains. Since no similarity was expected here, it seems that this is a better result than that produced by FASTA.

In this case, FASTA would have performed better with higher gap penalties, and this one case does not give sufficient evidence for any real comparison of the performance of the two methods. However, it does illustrate an important general point. The list of significant similarities ranked by *E* value is useful, but it is almost always necessary to inspect the sequence alignments to see where the sequences are most similar and where they are unrelated.

Sequence Filters

Nonspecific Sequence Similarity

Sequences or sequence segments may be similar to each other for reasons that are not evolutionary or do not reflect similarity in biological function. Sequences with biased composition or containing short period repeats are perhaps the most obvious examples of this. In protein sequences, it is common to find short segments whose composition is dominated by a small number of amino acids, and these segments are said to have low compositional complexity. For example, part of the sequence of the human Huntington's disease protein with *low complexity regions* associated with compositional bias towards glutamine and proline. It is interesting that in this case the length of the low complexity glutamine repeat, produced by unstable trinucleotide repeat expansion, is related to the severity of the disease.

It is possible that two sequences each contain low complexity segments with similar composition (e.g. another protein containing repeated glutamine and/or proline), but this type of similarity is unlikely to indicate that the proteins are homologous or that they share similar biological function. It is much more likely to reflect a common mechanism for the creation of the low complexity or repeat region, such as unstable expansion of a trinucleotide repeat. Another type of nonspecific sequence similarity is that associated with proteins that form *coiled coil structures*. These structures comprise two alpha helices coiled around each other and are associated with a seven-residue pseudo-repeat in the amino acid sequence, in which residues one and four are predominantly nonpolar. Examples of such structures are found in keratin and myosin. It seems that this type of structure is one that is advantageous to peptide sequences in general, and has been formed by many nonhomologous proteins in a process of convergent evolution to a similar structure. The presence of a shared coiled coil structure is not strong evidence for an evolutionary relationship between two proteins.

Similarity Searches

Similarity searches usually aim to detect homologous proteins or proteins with similar functions. Nonspecific sequence similarity therefore degrades the results of such searches. It is common to remove segments of the query sequence that might be associated with nonspecific sequence similarity prior to carrying out a similarity search. Several programs (which are linked from the text WWW site) are available to carry this out. The programs *SEG* and *DUST* will detect low complexity sequences. SEG is used mainly for protein sequences but will work with nucleic acids, and DUST is exclusively for nucleic acid sequences.

The program *XNU* will detect short periodicity repeats. The program COILS will predict potential coiled coil structures in proteins. Many WWW based sequence similarity search servers give the user an option to use the standard filters (SEG, XNU or DUST) prior to carrying out the search. These typically take the query sequence and replace monomers in the identified regions with the letter *X* so that they are ignored by the search software.

Iterative Database Searches and PSI-BLAST

Detection of Evolutionary Relationships

Divergent evolution changes the sequences of homologous proteins. In long evolutionary times, these changes can be very significant, and the degree of similarity between the homologous sequences can become very small. In many cases, the degree of sequence similarity falls below the level that is detectable by methods like BLAST and FASTA. This happens when the similarity between the homologous protein sequences is similar to the level of similarity that commonly occurs by chance between unrelated sequences. In the case of proteins, we can tell that this has happened by looking at the protein three dimensional structures, because, structural similarity tends to be preserved even when the sequences have diverged beyond recognition.

Well-known examples of this are found within the globin family of homologous proteins. These proteins share a common function (oxygen transport) and have the same structure, and yet there are pairs of members of the family from different organisms whose sequences have less than 10% identical residues. Using the database of three-dimensional protein structures, it has been estimated that BLAST and FASTA may detect only 20% of the total number of *distant* evolutionary relationships to the query sequence within the database. Therefore, much bioinformatics research effort has been devoted to methods that are able to improve on this detection rate.

Iterative Database Searches

PSI-BLAST makes use of iterated BLAST searches in order to extend the number of evolutionary relationships detected. An initial BLAST search is performed with the query sequence. Then, each of the hits from this search (above a chosen E value cutoff) can be used as the query sequence in a second iteration of BLAST searching. This second iteration should detect a greater number of evolutionary relationships to the initial query sequence, and this process can be repeated (iterated) until no more significant sequence similarities are found. After a number of iterations, more of the distant evolutionary relationships to the initial query sequence should have been detected. This is achieved by using information from *intermediate* sequences.

The evolutionary relationship between sequences A and C may not be detectable by BLAST, but if an evolutionary relationship between A and an intermediate sequence B can be detected, along with a further relationship between B and C, then it may be inferred that A is related to C. PSI-BLAST works in essentially this way. The only difference from the explanation above is a minor technical one. The second and subsequent iterations are not carried out as BLAST searches with each single sequence hit from the previous iteration, but rather as a BLAST search with a *sequence profile* formed from all the hits.

Sequence profiles are probabilistic models of sets of related sequences that contain information from all the sequences. Using

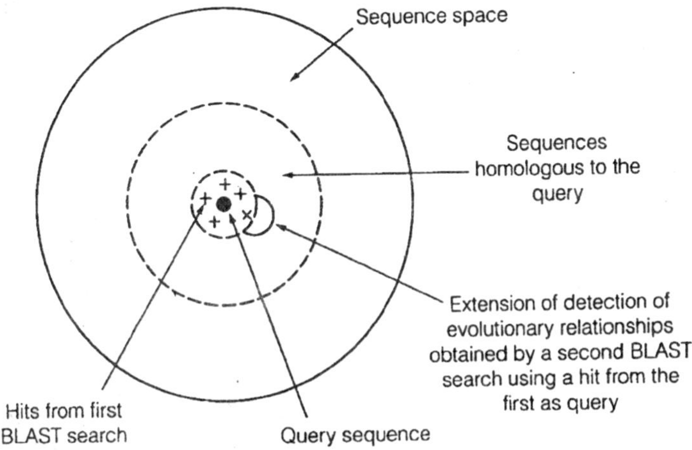

Fig. 4.6. The idea behind PSI-BLAST.

sequence profiles means that only one BLAST search is required for each iteration, rather than a BLAST search for each of the new hit sequences from the previous iteration. In practice, PSI-BLAST searches often detect up to twice as many evolutionarily related sequences as ordinary BLAST searches, which makes it a very useful tool indeed. There is one potential problem with iterative database searches. This is contamination with unrelated sequences. If the hits from anyone iteration of BLAST contain a false positive then this sequence will contaminate all further iteration with its own relatives that are not relatives of the original query. For this reason the *E* value cutoff used to choose the sequences to be employed at the next iteration of PSI-BLAST is chosen by default to be conservative, in order to minimize the risk of a false positive entering the search. A related problem that the internal workings of PSI-BLAST have to contend with is the domain structures of proteins.

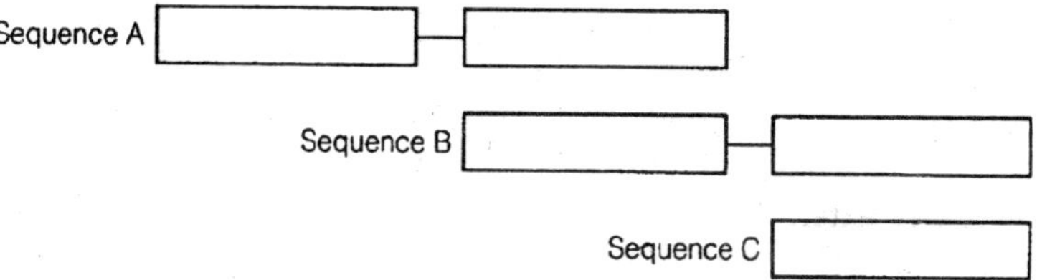

Fig. 4.7. Shared domain can mean that an intermediate sequence B does not relate sequence A to sequence C.

It is possible that an intermediate sequence (sequence B above) could be related to A by sharing a domain and to C by sharing a different domain. In such a case, there is no reason to infer that A is related to C. A PSI-BLAST server can be found at the National Center for Biotechnology Information (NCBI) and is linked from the text WWW site. Table 4.2 shows an example of the use of PSI-BLAST.

Table 4.2. Detection of PH domain sequences at progressive PSI-BLAST iterations

Iteration number	*Number of PH domain sequences found*
1	93
2	607
3	622
4	622

In this case, the query sequence was the first pleckstrin homology (PH) domain from human pleckstrin and the objective of the search was to find other PH domains. These domains are known to be very common, but to have very divergent sequences. The parameters for the search considered a hit significant if its E value was below 0.01, and hits with E values below 0.005 were used to form the sequence profile at each iteration. While the first iteration (an ordinary BLAST search) found only 93 hits, subsequent iterations added more than 500 new sequences. The search converged (no more new sequences were found) at the fourth iteration. This test case shows conclusively the advantage of PSI-BLAST over an ordinary BLAST search in detecting distant sequence similarity.

Vector Seeds: An Extension to Spaced Seeds Allows Substantial Improvements in Sensitivity and Specificity

We study techniques for faster and more sensitive pairwise local alignment. Recent advances have demonstrated modifications of the basic approach introduced in BLAST that lead to significant improvements in both sensitivity and running time of local alignment. Here, we present a framework unifying and further extending these approaches, leading to even better performance. The traditional approach for fast local alignment problem is represented by the BLASTN program.

BLASTN first identifies all pairs of short exact matches between the two sequences (hits). Every hit is then extended to a longer alignment, and alignments with high scores are reported, while those with low scores are discarded. High scoring alignments that do not contain a hit cannot be found by this approach. *Sensitivity* can be increased by decreasing the required length of the hit; however, this also increases the number of spurious hits (decreasing *specificity*) and thus also increases the running time. Thus there is a trade-off between sensitivity and specificity induced by particular definition of a hit.

Recently, several researchers have reported alterations to the hit definition that improve sensitivity without decreasing specificity. Kent in his program BLAT allows a *fixed number of mismatches* in the region that makes up a hit. For example, we may require at least 11 matches in a region of length 12. PatternHunter uses so called *spaced seeds*, which allow arbitrary numbers of mismatches in fixed positions of the hit. For example, a region of length 9 of an alignment is a hit to the PatternHunter seed 110110001 if there is a match on the first, second, fourth, fifth and ninth positions of the region. The other positions are not relevant. Ma et al. also introduce the idea of *optimizing the seed*, i.e. choosing the seed with given specificity which has the highest sensitivity under a given probabilistic model of alignment (region with 70% similarity of length 64).

In follow-up work, more realistic probabilistic models of alignments have been shown to yield optimal seeds with better performance on real data. (For our purpose, the performance of a seed is its sensitivity and specificity.) These methods also allow to create seeds tailored for particular application such as finding homologous protein coding regions. Protein alignments are scored by substitution matrices such as BLOSUM62 that define different scores for different matching and mismatching amino acid pairs.

Therefore techniques considering only matches and mismatches for finding hits do not work very well. BLASTP defines a hit as several consecutive positions with total score exceeding a given *threshold*. These techniques are often supplemented by requiring *two non-overlapping hits* that are evenly spaced in both sequences. This method also increases sensitivity compared to a single stronger hit. Here, we generalize these approaches into a new model of *vector seeds*.

Our model is general, allowing us to produce seeds for nucleotide alignments that incorporate positions that are required to match, positions that are free to vary, sets of positions that allow for a maximum number of variation, and more. Moreover, vector seeds allow easy transfer of techniques developed specifically for match/mismatch models into models based on scoring matrices. For example, we can apply spaced seed ideas to protein homology search. We also provide an algorithms to predict the sensitivity of a vector seed, given a probabilistic model of alignments. Our algorithm extends the original algorithm used to compute the sensitivity of PatternHunter's spaced seeds to the case of vector seeds. We have previously extended this algorithm to more realistic probabilistic models of alignments, such as HMMs, and the sensitivity of our new seed models can also be computed for HMMs.

This algorithm can be used to find the seed with the best predicted sensitivity for a given family of factor seeds. We further extend the algorithm to allow it to predict the sensitivity of two-hit alignment methods. Our extensions universally allow greater sensitivity and specificity over existing pair wise alignment methods. For coding DNA alignments, we greatly improve over the performance of BLAT or PatternHunter seeds specifically chosen for this problem, allowing false positive rate several times smaller than previously existed, or offering large advantages in specificity with comparable sensitivity. Our methods offer substantially improved performance over BLAT or Pattern-Hunter, with minimal additional required changes.

Alignments and Vector Seeds

Vector seeds are a new way of defining a *hit*, the conserved part of an alignment that triggers alignment extension in the homology search program. A good definition of hit allows efficient identification of all hits in a sequence database and leads to high sensitivity and specificity. In this section we introduce vector seeds as well as probabilistic models for predicting their performance.

Vector seeds

To define a hit in the vector seed model, we present ungapped pairwise local alignments as a sequence of real numbers, each corresponding to a position in the alignment. Alignments may potentially contain gaps. However, the hit must be located inside a single ungapped region of an alignment. Here, we model only individual ungapped fragments of such alignments. In the simplest case, we represent pairwise alignments as binary sequences. Zero represents a mismatch and one a match. For protein alignments, we represent the alignment between sequences $Y = y_1 y_2 ... y_n$ and $Z = z_1 z_2 ... z_n$ by the sequence of positional scores $(s_{y1,z1}, s_{y2,z2}, ..., s_{yn,zn})$, where $S = (s_{i,j})$ is the scoring matrix. We call such sequence of positional scores an *alignment sequence*. Now we are ready to formally define a vector seed.

Definition. A vector seed is an ordered pair: $Q = (v, T)$, where v is the seed vector $(v_1, v_2, ... v_l)$ of real numbers and T is the seed threshold value.

An alignment sequence $X = (x_1, x_2, ..., x_n)$ hits the seed Q at position p if $\Sigma^l_{i=1} (v_i \cdot x_{p+i-1}) \geq T$. That is, the dot product of the seed vector and the alignment sequence of length l beginning at position p is at least the threshold T. The number of nonzero positions in the vector v is the support of the seed.

Vector seeds generalize the spaced seeds of PatternHunter, the mismatching seeds of BLAT, and the minimum word score seeds used by BLASTP. For example, the BLASTP. For example, the BLAT seed that requires seven matching positions in nine consecutive positions in a nucleotide alignment is the vector seed ((1, 1, 1, 1, 1, 1, 1, 1, 1), 7). The PatternHunter seed 110110001 can be represented as the vector seed ((1, 1, 0, 1, 1, 0, 0, 0, 1), 5). The BLASTP rule that a hit is three consecutive positions having total score at least 13 corresponds to the vector seed ((1, 1, 1), 13). However, vector seeds can also encode more complicated concepts. For example, if the alignment sequence is binary, the vector seed ((1, 2, 0 , 1 2, 0, 1, 2), 8) requires matches in all positions with seed vector value of two, but allows one mismatch in the three positions with value one.

The positions with value zero are not relevant to a hit. Or, if the alignment sequence is over the values {0, 1, 10}, then the seed ((10, 10, 1, 10), 301) matches either the alignment vector (10, 10, 1, 10) or the vector (10, 10, 10, 10) but no others. Of course more complicated vector seeds than these rather simple example could be developed; the framework is general enough to allow vector seeds matching any half-space of R^l. However, for simplicity we will focus on a few families of vector seeds: for nucleotides, we will consider seed vectors with only zeros, ones and twos, where the total number of allowed mismatches is at most one or two, while for amino acids, we will consider short binary seed vectors.

Vector seeds are not universally expressive. For example the seed ((1, 1, 0, 1, 1, 0, 1, 1, 0, 1, 1)8), corresponds to requiring matches in the first two positions of each of our consecutive codons. There is no way in the vector seed model to encode the requirement that three of the four codons are matched this way; the seed ((1, 1, 0, 1, 1, 0, 1, 1, 0, 1, 1), 6) also allows one mismatch each in two codons.

Identifying hits in a sequence database

Assume we are given two sequences (or sequence databases) and want to find all hits between them. If hits are required to be exact matches of length k, the common approach is to create a hash table of all k-mers in one of the sequence and then search for each k-mer of the other sequence in the table. If hits are not exact matches (such as in BLAT or BLASTP), we can take each k-mer in the second sequence, generate a list of k-mers that would produce a hit and search for these k-mers in the hash table. This approach extends to the vector seed scenario.

Notice that we need to hash only characters on positions corresponding to non-zero elements in the vector seed. Hence, we seek vector seeds with small support that allow for a small number of hash table entries to be examined for each position in a query sequence. Otherwise, our results would not be of practical use! In particular, if most examined hash table entries are expected to be empty, the time to find the false positive will dominate the time required to verify whether seed hits are false positives; this is undesirable. For example, all alignments of at least 50% identity and length at least 100 contain a hit to the vector seed with vector (1, 1, ..., 1) of length 100 and with threshold 50. This seed will also be useless, as we would have to find matches to each 100 base pair sequence, and searching for such sequences could not be done efficiently.

Seed probabilities

So far we have described how to use a single vector seed to generate a set of hits. However, given a family of seeds, we desire the seed that will perform best. To allow such optimization, we represent properties of alignments by a probabilistic model and search for a seed maximizing probability of at least one hit in an alignment samples from the model. However, to control runtime, one must also control the false positive rate. Given two probabilistic models, one modeling alignments of unrelated sequences, and one for true alignments, we seek seeds with high sensitivity to true alignments, and low false positives.

Probabilistic models of true alignments

We model gap-free local alignments with probabilistic processes that generate sequences of real numbers. We investigate three models for local alignments. The simplest is the model introduced by PatternHunter for nucleotide alignments, with alignments of length 100 and each position having probability 0.3 of being zero (mismatch) and 0.7 of being one (match), independently.

The other model we use for nucleotide alignments is a three-periodic model of alignments in protein coding regions, where each triplet is emitted as a unit chosen from a probability distribution over $\{0,1\}^3$. Each triplet is independent of the others in this model. Such models can be used to effectively model the conservation in coding alignments, which are of key importance. Recently, we have shown that the optimal spaced seed for coding alignments is quite different from the one that optimizes PatternHunter's model, and showed that this simple codon model is moderately effective at representing these alignments.

To estimate seed sensitivity, we represent this probabilistic model as an 8-state hidden Markov model emitting individual binary characters. For protein sequences, we represent the alignment as a sequence of BLOSUM62 scores ranging from –4 to 11, and we use a positionally independent model similar to PatternHunter's. An amino acid scoring matrix implies a probability distribution on a pairs of residues being aligned in true alignments.

In particular, an entry $s_{i,j}$ in a scoring matrix implies that the probability of residue i aligning with residue j in related sequences is approximately $b^{s_{i,j}}$ times the probability of them aligning by chance, for some base b, and we use this observation to compute a distribution on positional scores implied by the matrix. Alignments are typically made up of more than one ungapped fragment, separated by gaps, and in our experiments, we treat the number and lengths of these fragments as random variables. Assume we know the probability $P_Q(l)$ that an ungapped alignment of a fixed length l, generated by a probabilistic model M, has a match to a given vector seed Q. These probabilities can be used to compute the probability that a seed matches an alignment sequence whose length is a random variable L, by simply computing $P_Q(L) = \Sigma_l p_Q(l)\ \Pr[L = l]$.

We have noted that alignments between homologous proteins usually consist of one long ungapped fragment and some number of shorter fragments. If the lengths of the long and short fragments are known distributions L and S, and the number of short fragments is random variable F, we can compute the probability that an alignment has any matches:

$$1-(1-P_Q(L))\left(\sum_{n=0}^{\infty}(1-P_Q(S))^n \Pr[F=n]\right).$$

Background model

To control the false positive rate, we need to be ale to computer the probability p that a hit of a given seed occurs purely by chance at a given pair of unrelated sequence positions. The expected number of spurious hits is then *pnm* where *n* and *m* are the length of the two sequences. For both nucleotide alignment models, our background probability distribution is a simple noise model, with zero emitted with probability 0.75, and one with probability 0.25.

For protein alignments, we use the probabilities of the background distribution built into the BLOSUM62 matrix to determine the probability that residues are aligned by chance, and then use that to compute the score distribution. Given this background probability, we can either use the algorithms given in the next section to compute the probability of a random match between two sequence, assuming that the sequence length is *l*, or, if the seed is particularly simple, we can just compute it directly. For example, in the background model for nucleotide alignments, the probability of a 0.1 seed with support s and threshold *T* having a match at a random place is just

$$\sum_{k=T}^{s} \binom{s}{k} 0.25^k \, 0.75^{s-k}.$$

Computing Sensitivity for Vector Seeds

Here, we show how to compute the sensitivity of a vector seed to detect alignment that come from a position-independent alignment model. Out method is analogous to the original Keich et al. algorithm, except that the alphabet has chanted and need not be binary, and that the definition of a hit is the more complicated dot product property. In recent work, we show how to extend the original Keich et al. algorithm to the case where the alignment sequence is generated by a hidden Markov model.

The extension to HMMs for vector seeds is straightforward, and we omit it for brevity. Suppose that the probabilistic model generates an alignment sequence $X = (X_1, x_2, ..., x_n)$, and that the value in each position is chosen from a small finite set D of real numbers. For each $d \in D$, $\Pr[x_i = d] = p_d$, and all positions are independent. We seek the probability that sequence X generated by this process has a hit of a given vector seed $Q = (v, T)$, where $|v| = l$. Let D^* be the set of all sequences of number from D whose length is at most l.

In the analysis of our algorithm's runtime, we assume that e can represent vectors from D^* as integers and manipulate them in constant time. This assumption is reasonable in our experimental setting. We compute the probability by dynamic programming, where the sub problem is the probability $P_Q(k, Z)$ that a sequence of length k, which begins with a given sequence Z from D^*, hits the seed Q. We are looking for $P_Q(n, \lambda)$, the probability of a hit when we make no conditions on the string, and the string is of length n. (We use λ to denote the empty sequence.)

We first identify sequences which are *guaranteed hits* and *possible hits*. Let F be a set of all sequences Z from D^* for which all extensions Z' to sequences from D^l have $Z' \cdot v \geq T$. (That is, all extensions are a hit of the seed). Let M be a set of all sequences Z from D^* for which there exists an extension Z' from D^l with $Z' \cdot v \geq T$ (Members of Z can be extended to seed hits.) With these definitions, $P_Q(k, Z)$ can be computed as follows:

1. If $k < l$, then $P_Q(k, Z) = 0$.
2. Otherwise, if $Z \in F$, then $P_Q(k, z) = 1$.
3. Otherwise, if $Z \notin M$, then $P_Q(k, Z) = P_Q(k-1, Y)$, where $Z = z_1 \ldots z_m$ and $Y = z_2 \ldots z_m$.
4. Otherwise, $P_Q(k,Z) = \sum_{d \in D} p_d P_Q(k, Z \cdot d)$.

The computation can be rearranged so that the third case is never reached, by always shifting forward enough positions when d is added to the end of Z in the fourth case. We move to the longest suffix of $Z \cdot d$ that is in M, and skip each instance of the third case. For each entry in M, and each value in D, this skip value can easily be computed in $O(l)$ time. Sets F and M can be found in $O(|M|)$ time. We initially compute the best and worst possible score for each suffix of the seed Q. Initially, F is empty and M contains the empty string.

In each iteration we choose one string X from M and using the pre-computed scores we identify which elements d from D permit or require an extension of Xd to a sequence of length l to hit the seed. Thus for an arbitrary vector seed Q, the algorithm computes $P_Q(n, \lambda)$ in $O(|M|(l + |D|n))$ time, if the entries of the dynamic programming table are stored in a data structure with $O(1)$ access. Note that the running time is dependent on the size of M; however, seeds with many matching strings are less useful in practice. Also, we need only keep the table P_Q for l values of k, so the memory requirement is $O(|M|l)$.

Some algorithm extensions

We note four other fairly straight forward extensions. The first is to seed pairs, where two seeds, Q_1 and Q_2, must both have a hit in the sequence. We first compute the P_{Q1} and P_{Q2}, matrices, and then, when we reach a prefix Z that is in the set F of guaranteed hits for either seed, we keep the prefix Z, and switch to requiring a hit to the other seed. The overall run time is at most twice the runtime to compute P_{Q1} and P_{Q2}. We can also compute the probabilities when a hit of either of the two seeds is required, or expand the set of matching strings in a variety of ways, by simply changing the sets M and F. For example, one can easily examine a single seed and the BLAST-style vector seed $(1^k, k)$, by simply adding the sequence 1^k to F and all of its $k - 1$ prefixes to M. We an also consider multi-hit models.

Here, we require that the alignment contains a least p hits at least l positions apart. This can be incorporated by keeping matrices $P_{Q,a}$, where $P_{Q,a}$ (k, Z) is the probability of at least a hits in a sequence of length k starting with Z. The recurrence is the as before, except for the following modifications. First, $P_{Q,1}$ $(k, X) = P_Q$ (k, X). Second, if $Z \in F$ and $a > 1$, then $P_{Q,a}$ $(k, Z) = P_{Q,a-1}(k - l, \lambda)$. This is because if a sequence of length k starts with a hit, we need $a - 1$ hits in the rest of the sequence. WE investigate the theoretical properties of two-hits in the rest of the sequence. We investigate the theoretical properties of two-hit BLAST and its variations, for both protein and nucleotide models.

We also study two-hit vector seeds in general, comparing their sensitivity and specificity to one-hit models. Finally, the probabilities p_d also need not be the same for all positions, as long as positions are independent. One can instead incorporate a position-specific probability distribution on D. This is equivalent to computing the sensitivity of a seed to a position-specific score matrix, if we assume that the true positives are generated with the probabilities implied by the score matrix. Of course, such scoring matrices are simply special cases of HMMs, which our algorithms can also expand to cover, using the techniques in our recent paper but the algorithm is especially straight forward for these profiles.

Experiments

We performed four experiments to verity the usefulness of vector seeds. Our first two experiments investigate their predicted performance in the simple Pattern-Hunter model of alignments. In one case, we compute the best single seeds, and in the other case, we study pairs of seeds that join together well.

In our other experiments, we used models trained for DNA coding regions and for proteins, and we computed both theoretical performance of seeds and actual performance on real sequences. In each experiment, we computed the probability of one hit and of two hits for the seeds we considered. We also computed the false positive probabilities for thee seeds (assuming that two-hit models) were satisfied when two matches occurred within 100 position of each other). In experiments on real data, we also computed how many alignments from our set can be detected by each seed.

Predicted Performance in the PatternHunter model

One hit required

First, we studied all vector seeds, (v, T) with vector entries zero or one, support s satisfying $8 \leq s \leq 15$, threshold T satisfying $s - 2 \leq T \leq s$, and whose length is at most min $\{s + 4, 17\}$. We HAVE Evaluated sensitivity of these seeds in a simple PatternHunter model. Seeds with both permitted mismatches and the structure of spaced seeds have a large advantage over either alone. For example, the no-mismatch seed PH-10 has false negative rate 22.4% and false positive rate 9.54×10^{-7}. By contrast, the two-mismatch vector seed VS-13-15 has false negative rate 4.11% with false positive rate 9.23×10^{-7}. This sed may not be practical, as there are 4^{15} possible hash table entries, but the more practical one mismatch seed VS-11-12 has false negative rate 4.9%, with twice the false positive rate (2.2×10^{-6}). This is to be compared to BLAT-11-12 with roughly the same specificity, but much lower sensitivity (almost three times as many false negatives).

Spaced seeds permitting errors are much more useful than unspaced seeds allowing errors. For example, the one-mismatch seed BLAT-11-12, with one hit, has false negative rate 14.8% while the best vector seed, VS-11-12, has false negative rate 4.9%, three times lower. Both have the same false positive rate, and are equally simple to implement.

Two hits of the same seed required

The situation is even more dramatic if we may require two hits in the alignment. For example, the seed VS-9-11, allowing two mismatches, has unacceptably high false positive rate for one hit. If we require two hits, however the false positive rate drops to 1.58×10^{-6}, comparable that for one hit to the VS-11-12 seed. Yet the false negative rate is an astonishing 0.1%. While there is some overhead involved in throwing out the many single hits that are not extended, this can still be done extremely quickly.

If one seeks much better specificity, VS-11-12, with two hits, allows false positive rate 4.8×10^{-10}, over three orders of magnitude better than for one hit to PH-10, with comparable false negative rate (19.1%).

The vector seeds with support between ten and twelve may be appropriate for practice with two hit models. If an input sequence is large, say 20 Mb, the expected number of entries in each site of a hash table will be at least one for sees of these supports, and the added work to identify double hits should be moderate for these seeds. Our tentative recommendation is two hits to the seed VS-11-12, with false negative rate 19.1% (better than one hit to the best seed of support 10), and false positive rate 4.86×10^{-10}.

Interestingly, we found that the sensitivity of a seed to one hit is an excellent predictor of its sensitivity to two hits. The sensitivity of a seed to two hits is quite consistently close to the cube of the sensitivity to one hit, with correlation coefficient $r^2 = .9982$. The fourteen seeds of support 11 allowing no errors with highest sensitivity to one hit are also the best for sensitivity to two hits; this pattern is consistent for other supports and seed lengths. This suggests that one need only consider sensitivity of seeds to one hit, perhaps computing sensitivity to two hits for appealing seeds.

Two seeds are better than one

One can use a pair of seeds instead of one, allowing matches to either seed. We can avoid twice completing alignments that hit both seeds, so runtime will roughly double for false positives and not change at all for true positives found with both seeds. We considered adding a different seed of support 13 with threshold 12 to the seed VS-12-13, which has false negative rate 10.4% by itself.

The results are shown in Table 4.3. The best pair halves the false negative rate while the worst augmentation (non spaced sed with one mismatch allowed) only improves false negatives slightly, yet will still double runtime. Interestingly, one of the best seeds to augment the seed with is its mirror image. This seed has the same sensitivity by itself as V-12-13.

Table 4.3. Theoretical performance of two-seed models when a seed of support 13 and threshold 12 is added to the seed VS-12-13.

Seed vector	*False Negative*	*False Positive*
No augmentation	10.4%	5.96×10^{-7}
1111111111111 (worst)	8.7%	1.18×10^{-6}
1110010111111111 (median)	6.1%	1.16×10^{-6}
1111010110111011 (mirror)	5.6%	1.18×10^{-6}
1110111011010111 (best)	5.4%	1.16×10^{-6}

Of course, there is no evidence that the best seed pair includes the best solo seed in it; however, it is a reasonable heuristic. (The best pair found here was also superior to 1000 random pairs of seeds). It is also sensible in the context that one is unhappy with the results of a search merely using the first seed.

DNA Seeds for Coding Alignments

We conduced further experiments on a data set consisting of alignments of homologous coding regions from human and fruit fly. The initial set contained 339 protein alignments. One protein alignment can yield several DNA alignments (or *fragments*) because the coding regions for each protein can be interrupted by non-coding introns. The final data set contained 972 un-gapped fragments in the training set and 810 gapped fragments in the testing set, after discarding weak and short fragments. A detailed description of the data set can be found elsewhere and the data set can be obtained at http://www.bioinformatics.unwaterloo.ca/supplements/03seeds. We used the training set of estimate the probability that a codon triplet has a given alignment pattern and the alignment length distribution. First we have investigated 1372 binary vector seeds (v, T) with supports $s \in \{10, ..., 15\}$ and $T \geq s - 2$.

The seeds we have investigated have codon structure: they can be divided into triplets, where each triplet is either (0,1,0), (1,1,0) or (0,0,0). In real alignments, the second codon position is most likely to be conserved and the third often varies. We compared the theoretical performance of the codon-aware spaced seeds versus their performance on our test set. Here, we also found quite striking advantages of vector seeds over seeds that are unspaced or that do not allow error.

The model is a good predictor of the performance of a seed, though our seeds to better than predicted, as the model is not aware of highly conserved parts of alignments. We then chose the best seeds allowing mismatches, at each support/threshold combination, and explored the effect of fixing the middle positions of some of their codons, by setting the corresponding position in the seed vector to two and raising the threshold by one. It shows the performance of a collection of BLAST and BLAT seeds, optimized spaced seeds, and optimized vector seeds, chosen for sensitivity to this model. This experiment highlights the expression richness of vector seeds. The seed CVS2-15-15 matches fully 95% of alignments. This is comparable to the sensitivity of the BLAST-7 seed, whose false positive rate is ninety times higher.

It is also preferred over the BLAT-10-11 seed for both sensitivity and specificity, and over the CPH-10 seed, which has complete specificity. Lastly, we note that the seeds we examined are also overwhelmingly preferred for two-hit study as well. While they are long enough that they reduce sensitivity below 65% in two-hit models still better than the BLAST-9 seed, they also admit tens of thousands of times fewer false positives. One can also use shorter seeds for use with two hits, which still allow for simpler hash table structures. The last seed is especially appealing; the support for this seed is just eight, making for very simple hash table structures, while the performance is comparable to one hit to the longer seed CVS2-13-12 in the table.

Seeds for Protein Alignments

We also studied the use of vector seeds for protein alignments. Out data set consisted of randomly chosen BLASTP 2.0.2 alignments of pairs of human sequences, taken from 8654 human proteins in the SWISSPROT database, release 40.38. WE chose 566 alignments with score between 50 and 75 bits using the BLOSUM62 scoring matrix, with a requirement that there are at most eight alignments from each collection of genes connected by BLASTP hits. (This restriction avoids choosing too many matches from one family of very common proteins; without it, the majority of alignments would be from only one family). To train our probabilistic model we estimated the distribution of the number of ungapped fragments, the length of the longest ungapped fragments, and the length of all other fragments as gamma distributions from this set of alignments. Other parameters of the probability distribution were obtained directly form BLOSUM62 matrix.

Table 4.4. Parameters characterizing the gamma distributions that approximate the properties of protein alignments from our data set.

Property	*N*	*mean*	*std. dev.*
Number of fragments	566	5.80	4.23
Length of longest fragment	566	59.6	20.4
Length of other fragments	2718	20.4	13.1

We investigated 237 seed of vector length at most six with between three and five ones and the remaining positions all zero. We chose values of T between 11 and 18 for seeds with 3 ones, between 12 and 22 for seeds with 4 ones, and between 15 and 25 for seeds with 5 ones. There is an observational bias in our experiments since the test set consists of alignments found by two-hit BLASTP. This guarantees that the two-hit BLASTP seed will match all alignments! The advantage of spaced seeds over unspaced seeds is not as dramatic as for nucleotide matches.

This is because BLASTP sed matches for proteins are more independent than BLASTN hits are for nucleotides. For example, the probability of a pair of amino acids scoring at least +5 is only 0.22, in homologous protein sequence, while the probability of a 1 is 0.7 in the PatternHunter model. Thus, for proteins, immediately following a hit there is a lower probability of another hit. Also, a disadvantage of spaced seeds is that they offer fewer positions for a possible match; when ungapped fragments are as short as they are for our protein alignments (with fragments of length 20 being reasonably common), this reduction affects seed sensitivity.

We also performed an experiment to verify that the false positive rates predicted were close to those found in reality. We chose 100 proteins at random and built a hash table for each one-hit seed in table 5. We then counted the number of seed hits when we chose 100 other random proteins. For all seeds, the false positive rate was within 10% of what was predicted by the background model. Still, there is some advantage to the use of spaced seeds. The spaced seed ((1, 1, 1, 0, 1), 14) offer a theoretical advantage of 61% fewer false positives over the standard BLAST seed, ((1, 1, 1), 13), with slightly lower false negative rate; for that matter, the obvious spaced seed of support 3, ((1, 0, 1, 1), 13), is also preferred over the standard BLAST seed, with 27% fewer missed alignments (though the difference is small). The advantage of two-hit models is not as great as for nucleotides, either. One notable discovery is that seeds depending on more than three positions have greater sensitivity than those depending on just three.

Given that there are only 160,000 amino acid 4-mers, for large protein databases, most hash table entries would be populated if one used seeds depending on four positions. If one is willing to tolerate an error rate comparable to the theoretical performance of the one-hit BLAST seed of threshold 15 (approximately 3.4% false negatives), the one-hit seed (1, 0, 1, 1, 1), 16), offers comparable false positives, with one third the false negative rate (1.2%). Or, if one is willing to use seeds of support 5, the two-hit seed ((1, 1, 0, 1, 1, 1)), 15) offers 4.0% false positive and seven times fewer false positives in theory.

A data set not derived from BLASTP

To try to avoid the observation bias we noted before, we aligned sequences that are not reported as aligned by BLASTP, but which are connected by a sequence of alignments that are reported by BLAST. If BLAST defines a graph on the set of proteins, these are nonadjacent nodes in the same connected component. We aligned all such pairs from connected components of size at most 30, and discovered 396 alignments in our target score range, again using the BLOSUM62 scoring matrix. However, all match the one-hit BLASTP seed with threshold 13, and all but three match the two-hit BLASTP seed with threshold 11.

Presumably, all were incorrectly thrown out during one of BLASTP's many filtering steps. In this new data set, we do see some advantage to spaced seeds, especially among less sensitive seeds. For example, the spaced vector seed ((1, 1, 0, 1), 16) matches 381 (96.2%) of these alignments, while the unspaced seed ((1, 1, 1), 16) matches 367 (92.6%) of them. Similarly, the spaced support 4 seed ((1, 1, 0, 1, 1), 18) matches 382 alignments (96.4), while the unspaced seed ((1, 1, 1, 1), 18) matches 373 (94.2%). We found similar results for two hit models.

Predictions versus reality

Finally, the simple model of sequences allows good predictions of seed sensitivity. The model generally does a good job at predicting the sensitivity of the seeds, except for the most sensitivity seeds. This is expected, since our test set includes only 566 alignments. Our set is also subject to observation bias: the alignments we tested were found with BLASTP, which guarantees that they have a hit to the two-hit seed ((1, 1, 1), 11)!

Conclusion

We have presented an extension to previous models for hits that are used in large-scale local alignments. Our vector seeds offer a much wider vocabulary for seed matches than previously studied seeds. For example, the allow certain positions to be more important than others, they allow a fixed number of mismatches in some positions and an arbitrary number in others, and more. Our extensions to spaced seeds or seeds with constrained mismatches allow substantially improved pairwise local alignment, with vastly improved sensitivity. Especially with the coding sequence nucleotide alignments (possibly one of the most important large-scale local alignment problems), alignment programs using our seeds can reduce their false negative rates by over half, with no change to false positives over BLAT or PatternHunter.

We have also shown on algorithm that allows us to predict the sensitivity and specificity of a vector seed on probabilistic models of true and random alignments. This allows us to choose an optimal seed for a given alignment task. Our algorithms is an extension to the original Keich et al. algorithm for predicting seed sensitivity in simple models. Our method is practical as long as the number of alignment sequences matching a given seed is moderate. Extensions to our algorithm allow one to predict the sensitivity of a seed in multihit models, or when using multiple seeds. We show spaced seeds can be helpful for proteins as well.

The improvement are not as dramatic as for nucleotides, mostly because the seeds themselves are so short, yet they are still useful, and if one is willing to allow a moderate false positive rate, spaced seeds are strongly preferred over unspaced seeds. We also show the contexts under which two-hit models are preferred over one-hit models for proteins. Our results offer substantial improvement over the current state of the art, with minimal change required in coding.

Future work

Finally, we discuss a few extensions which we have only begun to consider. In PsI BLAST, alignment phases after the first are based on position-specific scoring matrices, which model the probabilities expected in sequence matching a profile. Much as with the standard BLASTP model of aligning sequences that comes from the scoring matrix, here as well, one may desire a seed that has a higher probability of matching the sequence than the usual $(1^k, T)$ consecutive seed.

It is impractical to compute the probabilities for thousands of seeds for each profile, yet it is quite reasonable to compute the match rates for each of a small set of diverse good seeds, and use the best of these seeds in that round. (One may also compute predicted false positive rates, using the standard background probabilities with the new scoring matrices.) We are also interested in whether the representation of vector seed hits as half-spaces in a lattice can help in optimizing them. Given a certain length, support and threshold, is it possible to find the best seed, even for a simple position-independent model, without using essentially exhaustive search? While the exhaustive search is possible for small seed families, for the vector seed models, this becomes absurd as the families of possible seeds grows to the millions.

Clearly, one could use heuristic methods. However, sometimes, there is a large difference between the best seed and seeds at the 99th percentile, so one would want a very good heuristic. Finally, we demonstrated that the vector seeds are not universally expressive. For example, there is no way in the vector seed model to require that three of the four codons are match in the first two positions each. Is it possible to extend vector seeds so that they are more expressive ?

Systematic Statistical Analysis of Ion Trap Tandem Mass Spectra in View of Peptide Scoring

Tandem mass spectrometry (MS/MS) combined with database searching has become central in proteomics projects. Such projects aim at discovering all or part of the protein is present in a certain biological tissues, e.g. tears or plasma. Before MS/MS can be applied, the complexity of the initial sample is reduced by proteins separation techniques like 2D-page or liquid chromatography (LC). The proteins of the resulting simpler samples are digested by an enzyme that cleaves the proteins at specific locations. Trypsin is frequently used for this purpose. MS/Ms analysis is performed on the digestion products, which are named peptides.

Alternatively, early digestion can be applied and peptide separation techniques used. In both cases, the peptides are positively ionized and fragmented individually and, finally, their masses as well s the masses of their fragments are measured. Such masses constitute a data set, the experimental MS/Ms identify the peptides by searching into a database of peptide sequences. By extension, this procedure allows to identify the proteins. MS/MS database searching usually involves the comparison of the experimental MS/MS spectrum with theoretical MS/MS spectra, computed form the peptide sequences found in the database.

A (peptide) score function or scoring scheme is used to rate the matching between theoretical and experimental spectra. The database peptide with the highest score is usually considered as the correct identification provided the score is high enough and/or significant enough. Clearly the availability of sensitive and selective score functions is essential to implement, reliable and automatic MS/MS protein identification systems. In we proposed a generic approach (OLAV) to design such score functions. This approach is based on standard signal detection techniques. In this paper we apply it to the LC electrospray ionization ion trap (LC-SEI-IT) mass spectra and we study the relative interest of various quantities we can compute when we compare theoretical and experimental spectra.

We finally select a combination of such quantities and establish the performance of the corresponding score function by performing large-scale computations. For reference purposes, we give results obtained with Mascot, a widely used commercial protein identification program and we report the performance we obtain on a generally available data set for which Sequest results have been published. Currently available protein identification systems can be classified into three categories: heuristic systems, systems based on a mathematical model and hybrid systems.

In the heuristic category there are well known commercial programs: Mascot, Sequest and SONAR MS/MS. Sequest and SONAR correlate theoretical and experimental spectra directly, without involving a model. Mascot includes a limited model as well as several heuristics intended to capture some properties related to signal intensity and consecutive fragment matches. Model-based systems use stochastic models to assess there liability of matches. In this category we find: MassLynx, SCOPE Probld and SHERENGA. SCOPE considers fragment matches as independent events and estimates a likelihood by assuming a Gaussian distribution of mass errors. MassLynx uses a Markov chain to estimate the correct match likelihood and to model consecutive fragment matches. Probld uses Bayesian techniques to estimate the probability a match is correct.

It integrates several elementary observations like peak intensities and simultaneous detection of fragments in several series. SHERENGA estimates a likelihood ratio by considering every fragment match as an independent Bernoulli random variables. Improves over SHERENGA by considering signal intensity and neutral losses. The hybrid category generally uses multivariate analysis techniques to filter the results returned by heuristic systems. The knowledge of which are the essential quantities to include in a score function is certainly beneficial to most of the approaches above. According to the relative performance of the various score functions we tested, the most important quantity to include in a score function is the probability to detect each ion type.

The next quantity is the intensity of detected fragments: intense fragment must match with probabilities depending on the ion type. Then different extra quantities improve performance: probability to detect a fragment depending on its amino acid composition, probability to observe consecutive fragment matches. By combining these quantities in a naive Bayesian classifier, we design a score function that has a false positive rate as low as 3% is the true positive rate is fixed at 95%. On data set, the false positive rate is inferior to 0.5%. It is difficult to compare peptide score functions without testing them on the same data sat.

As a matter of fact, MS/MS data are noisy and of variable precision. Hence, the absolute performance of a given score function may change depending on data set quality. According to our experience, the relative advantage of a score function compared to another one is generally stable from data sets to data sets. To allow reader to compare our results with their own experience, we report them by using an available data set or with a standard algorithm tested on the same set. We observe a strong advantage in favour of the approach we propose.

Mass Spectrometry Concepts

ESI ion trap instruments

Current peptide ionization methods that are common in proteomics include electrospray ionization (SI) and matrix assisted laser desorption ionization. Several technologies are also available for selecting and fragmenting the accelerated peptides, one of which is quadrupole ion trap (IT). IT represents a significant and growing portions of the mass spectrometers used in proteomics. The approach presented in is not specific to ESI-IT mass spectra. ESI produces positively charged ions, whose charge states are mainly two or three.

An IT instrument breaks peptides by low-energy collision-included dissociation. The fragmentation process yields several ion types (a, b, y), depending on the exact cleavage location. The proportion of each ion type produced changes with the MS/MS technology. Additionally, certain amino acids can loose one water (H_2O) or ammonia (NH_3) molecule. Consequently, fragment ion masses can be shifted by –18 Da and/or –17 Da. Every mass spectrometry instrument produces a signal that can be assumed to be continuous. Peak detection software is used for extracting peptide or fragment masses from this signal. The list of extracted masses is named a peak list. When we refer to a spectrum, we always refer to the corresponding peak list, which is the primary data for identification.

Matching theoretical and experimental spectra

We do not describe here how to compute theoretical spectra. It is sufficient to know that, given an amino acid sequence, there exit precise rules to compute the mass of every possible fragment of each ion type. The theoretical spectrum consists of the masses of fragments for selected set S of ion types. S is instrument technology dependent. S also depends on the peptide charge state. We name the comparison of an experimental spectrum with a theoretical spectrum a *match*.

A match can be either correct or random. From the match we compute several quantities that are then used by the score function. These quantities are modeled as random variables. It is convenient to represent them by a random vector E. The score function is intended to distinguish between correct and random matches. This problem can be viewed as an hypothesis testing problem. We propose to build score functions as log-likelihood ratios as this approach yields optimal decision rules, provided E probability distributions are known in the correct (H_1) and random (H_0) cases. In practice, we have to approximate these two distributions. Nevertheless, we believe that log-likelihood ratios are very effective for peptide scoring, which is confirmed by the high performance we achieve. We give other arguments to justify this point of view.

Statistical Modeling

Data set

We analyzed by proteomics two pools of 2.5 litres of plasma. One control pool and on e diseased pool (coronary artery disease), each containing roughly 50 selected patients. Multidimensional liquid chromatography was applied, yielding roughly 13000 fractions per pool, which were digested by trypsin and analyzed by mass spectrometry (LC-ESI-IT) using 40 Bruker Esquire 3000 instruments. The set of ion trap mass spectra we sue is made of 146808 correct matches, 33000 of which have been manually validated. The other matches have been automatically validated by a procedure, which, in addition to fixed thresholds, includes biological knowledge and statistic about the peptides that were validated manually.

There are 3329 singly charged peptides (436 distinct), 82415 doubly charged peptides (3039 distinct) and 61604 triply charged peptides (2920 distinct). Every performance reported in this paper is obtained by randomly selecting independent training and test sets, which sizes are 3000 and 5000 matches respectively (1000/2329 for charge 1). This procedure is repeated 5 times and the results averaged. We also checked that both model parameters and performance barely change from set to set. In order to validate one important hypothesis, we use another set of 1874 doubly charged peptides (73 distinct) and 4107 triply charged peptides (90 distinct). The spectra were acquired on 4 Bruker Esquire 3000 + instruments. We refer to this data set as data set

B. Data set has been generated by a ThermoFinnigan LCQ ion trap instrument.

Theoretical spectrum and neutral losses

As we mentioned above, certain amino acids may lose water or ammonia (A-so-called neutral loss). By considering the chemical structure of amino acids, we observe that Arg (R), Asn (N) and Gln (Q) may loose ammonia, and Ser (S) and Thr (T) may loose water. In order to break disulfur bonds, Cys (C) are modified. A common modification is S-carboxamidomethyl cysteines (Cys_CAM, + 57 Da) whose chemical structure suggests a potential loss of ammonia. In the data set we use, Cys are modified as Cys_CAMs. To able to compute realistic theoretical spectra, we check which of the amino acids above loose water or ammonia significantly. This point has been already considered by for doubly charged peptides and based on a much smaller data set. We follow a similar approach, i.e. we assume that every amino acid may loose water or ammonia and we compute the amino acid composition of matched ions a, by, y and a,b,y-17, 18. Finally, the amino acid compositions are compared to find significant enrichments. The results are shown in Table 4.5 and we decide to exclude Cys_CAM. Asn in kept although it is only significant for singly charged peptides. We checked that including it provides a small benefit for singly charged peptides without penalizing higher charge states.

Comparing peptide score functions

To supplement the peptide scores with p-values, we have introduced in a method for generating random matches and thus random scores. The general principle matches and thus random scores. The general principle is the following: given a match, we generate a fixed number of random peptide sequences, with appropriate masses and possible PTMs, by using a Markov chain. The random peptides are matched with the original spectrum to estimate the random score distribution. Now, to compare the relative performance of various score functions, we compute, for each correct match, the ratio of the correct match score with the best of 10000 random match scores.

Basic reference score function

We define a first score function L_1, which we will refer to as the minimal score function. Every potentially improved score

Table 4.5. Neutral loss statistics.

Ions	*CysC*	*Asn*	*Gln*	*Arg*	*Mean*	*std dev*	*Ions*	*Ser*	*Thr*	*Mean*	*Std dev*
					Singly charged peptides						
a-17	1.2	1.7	1.3	0.7	1.00	0.25	a-18	1.2	1.4	0.98	0.24
b-17	1.2	2.1	1.5	1.9	1.08	0.37	b-18	1.2	1.3	0.94	0.19
y-17	1.0	1.9	1.2	2.3	1.13	0.42	y-18	1.2	1.0	1.04	0.14
					Doubly charged peptides						
a-17	1.0	1.9	0.9	0.4	1.02	0.28	a-18	1.2	1.2	0.99	0.21
b-17	1.2	1.6	1.2	0.9	1.00	0.18	b-18	1.2	1.2	0.93	0.19
y-17	1.2	1.5	1.5	0.9	1.04	0.19	y-18	1.1	1.2	1.00	0.13
					Triply charged peptides						
b-17	1.0	1.4	0.9	0.9	1.00	0.20	b-18	1.4	1.1	0.95	0.20
y-17	0.8	1.5	1.3	0.8	0.99	0.23	y-18	1.1	1.6	0.94	0.28
b^{++}-17	0.9	1.0	1.0	1.0	0.99	0.05	b^{++}-18	1.1	1.1	0.98	0.06
y^{++}-17	1.0	1.0	1.0	0.8	0.99	0.11	y^{++}-18	1.0	1.2	0.99	0.13

function will be compared to this one. L_1 is a slight extension (charge state dependence) of a score function introduced in and it can be derived as follows. We assume that fragment matches (tolerance given) constitute independent events and the probability of these events depends on the ion type $\theta \in S$ and the peptide charge state z. We denote this probability $p_\theta(z)$. Now, let $s = a_1 \ldots a_n$ be a peptide sequence and a_i its amino acids. The probability of a correct match between s and an experimental spectrum is estimated by taking the product of $p_\theta(z)$ for every matched fragment and of $1 - p_\theta(z)$ for every unmatched fragment. The null-model is identical with random fragment match probabilities $r_\theta(z)$. We find

$$L_1 = \log\left(\prod_{i=1}^{n}\left[\prod_{\theta \in M(s,i)} \frac{p_\theta(z)}{r_\theta(z)} \prod_{\theta \in S(s,i)-M(s,i)} \frac{1-p_\theta(z)}{1-r_\theta(z)}\right]\right).$$

$S(s, i) \subset S$ is the set of ion types ending at amino acid a_i, $M(s, i) \subset S(s, i)$ is the set of ion types matching experimental fragment mass. $S(s, i)$ may be proper subset of S because certain ions are not always possible depending on the fragment last amino acid (neutral loss). $p_\theta(z)$, $\theta \in S$, are learnt from a set of correct matches. The probabilities of random fragment matches $r_\theta(z)$ are learnt from random peptides. We use relative entropy in bit $H_\theta(z) = p_\theta(z) \log_2 (p_\theta(z)/r_\theta(z))$ to measure the importance of each ion type. For z

= 1, 2, 3, we empirically determined a threshold that is a constant divided by the average peptide length given *z*. The performance of L_1 and the list of ion types selected are robust with respect to threshold variations. In decreasing order of $H_\theta(z)$, we use : (*z* = 1) y, b-18, b, y-17, b-17, (z = 2), y, b, b-18, b-17, y-18, y-17 and (z = 3) y, b^{++}, b, b^{++}-18, y^{++}, b^{++}-17, y^{++}-18, y-18, y^{++}-17, b-18.

Table 4.6. Performance comparison

	Change 1			Change 2			Change 3		
Function	1.0	1.2	1.4	1.0	1.2	1.4	1.0	1.2	1.4
L_1	75.0	56.2	42.1	97.4	94.7	90.4	96.4	94.4	91.9
L_{consec}	73.3	54.8	40.8	97.8	95.0	91.1	97.8	96.2	93.9
L_{intens}	79.5	57.5	40.9	97.8	95.3	91.2	97.7	95.8	93.5
$L_{1,\ class}$	73.2	57.0	46.5	97.6	95.2	91.9	96.4	94.7	91.8
$L_{i\ Class}$	76.6	57.5	42.7	97.8	94.9	90.8	97.5	96.0	93.1
L^*_1	66.5	47.7	36.7	83.9	79.3	75.0	82.7	77.8	74.0
L^*_{consec}	73.8	55.6	42.5	85.5	79.3	75.7	89.0	84.8	82.4
L_{intens}	71.1	48.5	35.2	83.9	78.9	76.3	87.0	83.7	79.4
$L^*_{1,\ class}$	65.5	50.5	41.9	87.8	84.2	81.2	84.2	80.7	76.6
$L^*_{i\ Class}$	67.7	48.5	36.8	84.9	79.6	75.7	86.3	83.0	77.6
L^B_1				98.4	96.5	93.7	98.7	94.7	85.9
L^B_{intens}				99.9	99.8	99.4	99.9	99.9	99.3
L^{B*}_1				88.8	83.1	79.8	95.4	85.1	82.8
L^{B*}_{intens}				99.2	98.1	96.5	98.9	98.9	97.7

Consecutive fragment matches

The score function L_1 is based on a very strong simplifying assumption; fragment matches are independent events. In theory (very simplified), if an amino acid in the peptide has been protonated, then successive fragments should also contain this protonation site and hence be detected. In reality, the protonation sites are not the same for every copy of a peptide and a probabilistic approach should be followed. A natural improvement of L_1 would be a model able to "reward" consecutive matches, while tolerating occasional gaps.

This can be achieved by using Markov chain (MC) or a hidden Markov model (HMM). In this perspective as observed in, it may be advantageous to unify several ion types in one generalized ion type to better capture the consecutive fragment match pattern. For instance, one may want to consider ion types b, b-

17, b-18, b^{++} as one general ion type B. We consider one MC and two HMMs, denote by L_2 the log-likelihood ratio of models for consecutive matches. Given a choice of model (MC or HMM), we define a new score function $L_{consec} = L_1 L_2$, $L_2 = \Pi_{\theta \in S'} L_{2,\theta'}$ where S', the set of generalized ion types, and $L_{2,\theta'}$ the corresponding log-likelihood rations.

For each peptide charge state, we test every combination of the models hmmA, hmmJ and mcA both for the alternative (H_1) and the null hypotheses (H_0), with orders n = 2,3,4 order (H_0 mode) $\leq$ order (H_1 model). That is 84 L_2 models in total. We empirically set S' to (z = 1) Y = {y}, (z = 2) B = {b, b-18, b-17}, Y = {y, y-18, y-17} and (z = 3) B = {b, b^{++}, b^{++} -18, b^{++} -17}, Y {y, y^{++}, y^{++}-18}. The parameters are learnt by expectation maximization. We found no unique combination that would dominate the other ones. Several combinations perform well and, as a general tendency, we have observed that HMMs have a slight advantage for the H_1-model, whereas MCs are sufficient for the H_0-model.

The performance of the various combinations is similar, therefore indicating the intrinsic importance of considering consecutive "no matter" the exact method. The performance of the best combination (z = 1: hmmA(3)/mcA(2), z = 2: hmmA(4)/ mcA(3), z = 3: hmmA(4)/hmmJ(2)) is shown in Table 7.4. In every case, the transition and emission probabilities nicely fit the model structures. For hmm J(2), Y ions and charge 2, we find the transition START to S_1 (probability 1), S_1 to S_1 (0.59), S_1 to S_2 (0.41), S_2 to S_2 (0.88), S_2 to S_1 (0.12) and emission S_1 ('m' with probability 0.05, 'f' 0.95), S_2 ('m' 0.97, 'f' 0.03)

Signal intensity

It is well known among the mass spectrometry community that different ion types have different typical signal intensity. In the case of tryptic peptides, C-terminal ion types (x, y, z) naturally produce more intense peaks. This is due to the basic tryptic cleavage sites (Lys, Arg), which facilitate protonation. And even report fragment relative length intensity dependence. Here we use a simple model that orders the experimental peaks by intensity and then split them into 5 bins. We obtain $L_{intens} = L_1$, L_3, $L_3 = \Pi_{\theta \in S''} L_{3,\theta'}$ S'', a set of ion types, and $L_{3,\theta'}$ the corresponding log-likelihood rations. By selecting ion types for their significance (relative entropy), we set S'' = to (z = 1), b, b-17, b-18, y, y-17, y-18, (z =2) b, y and (z = 3) b-17, b^{++}, y, y-17, y^{++}. The performance of L^{intens} is shown Table 4.6. Although signal intensity improves performance, we expected a more spectacular change.

By further investigating, we found a direct explanation for this disappointing result. Bruker peak detection software allows for exporting the n most intense peaks above noise level into the peak list. At the time we generated our main data set, n was set to 100. Ion types like –18, b-18, y-17, y-18, b^{++}, y^{++} are generally important for scoring, although their signal is much less intense than b or y signal. Now, given that longer peptides statistically have more protonation sites, it is clear that singly charged peptides are shorter. Therefore, $n = 100$ is sufficient to cover most of the fragment masses. On the other hand, it turns out that it is not sufficient to include enough fragment masses in the model L_{intens} for $z = 2,3$. We used the complementary data set B, which was generated with $n = 200$. We denote by $L^{B}{}_{intens}$ the model L_{intens} trained and tested on this set. The result shown in Table 4.6 nicely confirm the explanation.

Since the spectra in data set B were acquired on a different (better) instrument and the samples were made of purified proteins (not a biological sample), we repeat the performance of L_1 for reference purpose. In theory, L_3 includes L_1 and one could expect that L_3 performance is similar to L_{intens}. In practice, L_3 performance is very inferior to L_{intens} (less than 66% for a ratio of 1), even on data set B. The reason is that the pattern captured by L_1 is more or less always available, whereas the intensity pattern is more variable.

Amino acid dependence

Depending on their amino acid sequence, fragments may be more or less easily detected. The actual dependence involves several phenomena. A model making use of the whole fragment sequence would contain too many parameters. It is commonly accepted that the last amino acid of a fragment (cleavage site) plays a significant role in the above mentioned phenomena. We limit the number of model parameters by only considering the last amino acid of a fragment and by grouping them in classes.

Basic score revisited

We designed an improved version of L_1, which we name $L_{1,class}$, that uses parameters $p_\theta(z, c)$, $r_\theta(z, c)$, c a set of amino acids. We empirically determined the following amino acid classes (amino acids with similar probabilities): (N-tern ions) 'AFHILMVWY',

'CDEGNQST', 'KPR' and (C-term ions) 'HP', 'ACFIMDEGLN-•QSTVWY', 'KR'

Consecutive fragment matches revisited

It is possible to extend hmmA, hmmJ and mcA by replacing their states by one state per amino acid class to separate amino acids that inhibit fragmentation from amino acids that favour fragmentation. As we prefer to stay with simple and robust models, we have not implemented the amino acid dependent versions of hmmA, hmmJ and mcA.

Signal intensity revisited

The model can be extended as we did not $L_{1,class}$, thus obtaining a model $L_{i\ Class} = L_1 L_{3,class}$. $L_{i\ Class}$ performs much worse than L_{intens}, what we explain by the fact that, although, the last amino acid plays a major role in the fragment dissociation phenomena, it is no strong relation with signal intensity. Signal intensity is more a consequence of the ion type and the entire peptide sequence.

An Efficient Score

L_1 is significantly improved by considering signal intensity. Consecutive fragments matches as well as the amino acid dependent version of L_1 also improve the performance. Accordingly, we tested 4 combinations, which are $C_1 = L_{intens}$, $C_2 = L_{1,class}L_2$, $C_3 = L_{intens}L_3$ and $C_4 = L_{1class}L_2L_3$. The receiver operating characteristics (ROC) curves are obtained by setting a threshold on match p-values. The correct match p-values are computed by searching the peptides of our data set against a database of 15000 human proteins with variable Cys-CAM and oxidation (Met, His, Trp) modifications. The random match p-values are computed by searching against a database of 15000 random p-values are computed by searching against a database of 1500 random proteins with the same variable modifications and by taking the best match. The random protein sequences were obtained by training an order 3 MC on the human protein database. We observe that C_4 is the best combination.

Discussion

In this work we designed score functions by assuming the independence of many statistical quantities. The choice allowed us to rapidly design efficient score functions as naïve Bayesian classifier. This can be seen as preparative work to identify key contributors to successful peptide score functions in order to design future models, comprising fewer simplifying assumptions. We would that ion type probabilities, signal intensity and consecutive fragment matches are essential to a peptide score function. We omitted one aspect of a match that may be pertinent: the simultaneous detection of several ion types (neutral loss, complementary fragments). Peptides elution times may also provide additional information to select correct matches.

New approaches based on the mobile proton model are under development, and it should be possible to combine them with what we presented here. The general applicability of modern stochastic score functions has not been addressed here. In particular, two central questions might limit their value: the minimal size of the training set and performance robustness. These two points are addressed in and the results are positive. We described a genetic method for designing peptide score functions. We applied it systematically to a large and diverse MS data set (140000 + peptides) to design a series of models of increasing complexity.

By selecting an appropriate combination of models, we obtained a very efficient score function, which, given a true positive rate of 95% has a false positive rate as low as 3% (doubly charged peptides) or 2% (triply charged peptides). This is 3.5 to 5.8 times less than Mascot 1.7. On data set our false positive rate is less than 0.5% which is a 14-fold improvement compared. In Sequest performance is reported when used "traditionally", i.e. by setting thresholds on Xcorr and other quantities exported by Sequest. The smaller Sequest false positive rate reported is 2% with 59% true positive rate. We obtain a corresponding false positive rate of 0.004%, which is 50 times less. The highest Sequest true positive rate reported (threshold set 2) is 78% with 9% false which is 187 times less. Similar performance has been obtained on real samples by using Bruker Esquire 3000 + instruments. Such low false positive rates, the lowest ones ever reported to our best knowledge, combined with the competitive price and robustness of ion trap instruments, provide a highly appropriate technology platform in the perspective of the many academic and private large-scale proteomics projects currently emerging.

5

CONFORMATIONAL SAMPLING

The goal of conformational analysis is to shed light on conformational characteristics of flexible biomolecules and to gain insight into the relationship between their flexibility and their function. Because of the importance of this approach, conformational analysis plays a role in many computational projects ranging from computer-aided drug design to the analysis of molecular dynamics simulations and protein folding. In fact, most structure-based drug design projects today use conformational analysis techniques as part of their toolchest. Structure-based drug design a rational effort is applied to identifying potential drug molecules that bind favorably into a known three-dimensional (3D) binding site, the structure of which was determined through *X*-ray crystallography, NMR spectroscopy, or computer modeling. Because such an effort requires, among other things, structural compatibility between the drug candidate and the binding site, computational methods were developed to "dock" ligands into binding sites. These docking calculations are used for screening large virtual molecular libraries, saving both time and money. However, although docking is fairly straightforward with rigid molecules, it becomes significantly more complicated when flexible molecules are considered. This is because flexible molecules can adopt many different conformations, each of which may, in principle, lead to successful docking.

Although there are a few "flexible docking" approaches that account for flexibility during the docking process itself, most docking applications rely on conformational analysis to deal with this problem (*e.g.*, by generating a multitude of molecular conformations that are docked separately into the binding site). The importance of conformational analysis in the context of drug design extends beyond computational docking and screening. Conformational analysis is a major tool used to gain insight for future lead optimization. Furthermore, even when the 3D structure of the binding site is unknown, conformational analysis can yield insights into the structural characteristics of various drug candidates.

In a different context, conformational analysis is essential for the analysis of molecular dynamics simulations. The direct output of a molecular dynamics simulation is a set of conformations ("snapshots") that were saved along the trajectory. These conformations are subsequently analyzed in order to extract information about the system. However, if, during a long simulation, the molecule moves from one conformation class to another, averaging over the whole simulation is likely to be misleading. Conformational analysis allows one to first identify whether such drastic confor-mational transitions have occurred and then to focus the analysis on one group of conformations at a time.

In view of their importance it is not surprising that conformation sampling and analysis constitute a very active and innovative field of research that is relevant to biomo-lecules and inorganic molecular clusters alike. The following sections offer an introduction to the main methodologies that are used as part of a conformational analysis study. These are arranged according to the three main steps applied in such studies : (1) conformation sampling, (2) conformation optimization, and (3) conformational analysis.

Conformation Sampling

Conformation sampling is a process used to generate the collection of molecular conformations that will later be analyzed. Ideally, all locally stable conformations of the molecule should be accounted for in order for the conformational analysis to be complete. However, owing to the complexity of proteins and even fairly small peptides it is impractical to perform such an enumeration. The number of locally stable conformations increases so fast with the molecular size that the task of full enumeration becomes formidable. Even enumerating all possible (ϕ, ψ), conformations of a protein backbone rapidly becomes intractable. As a result, most conformational studies must rely on sampling techniques. The basic requirement from such sampling procedures is that the resulting conformational sample ("ensemble") will be representative of the system as a whole. This means that in most biomolecular studies a "canonical" ensemble, characterized by a constant temperature, is sought. Therefore sampling methods that were designed for canonical ensembles and that guarantee "detailed balance" are especially suitable for this task. Two such methods are high temperature molecular dynamics and Monte Carlo simulations. However, because of the complexity and volume of biomolecular conformational space, other sampling techniques, which do not adhere to the canonical ensemble constraint, are also often employed.

High Temperature Molecular Dynamics

Molecular dynamics simulations are among the most useful methods for sampling molecular conformational space. As the simulation proceeds, the classical trajectory that is traced is in fact a subset of the molecular conformations available to the molecule at that energy (for microcanonical simulations) or temperature (for canonical simulations). Assuming that the ergodic hypothesis holds, an infinitely long MD trajectory will cover all of conformational space. The problem with room temperature MD simulations is that a shorter finite-time trajectory is not likely to sample all of conformational space. Even a nanosecond MD trajectory will most likely be confined to limited regions of conformational space. The room temperature probability of crossing high energy barriers is often too small to be observed during a finite MD simulation.

A common solution that allows one to overcome the limited sampling by MD simulations at room temperature is simply to raise the temperature of the simulation. The additional kinetic energy available in a higher temperature simulation makes crossing high energy barriers more likely and ensures a broad sampling of conformational space. In raising the simulation temperature to 1000 K or more, one takes advantage of the fact that chemical bonds cannot break in most biomolecular force fields. Namely, the fact that bonds are modeled by a harmonic potential means that regardless of the simulation temperature

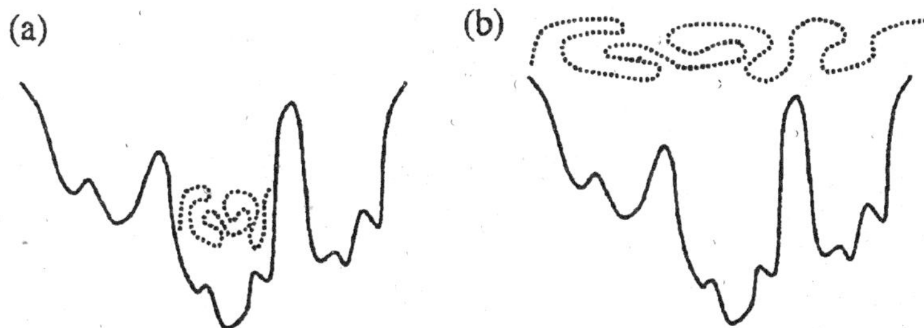

Fig. 5.1. A schematic view of (a) a low temperature simulation that is confined by high energy barriers to a small region of the energy landscape and (b) a high temperature simulation that can overcome those barriers and sample a larger portion of conformational space.

these bonds can never spontaneously break, and the chemical integrity of the molecule remains intact. The effect of the unrealistically high temperatures employed is primarily to "shake" the system and allow the molecule to cross high energy barriers.

There is no definite rule regarding what temperature is "high temperature" in this context, as this depends on the character of the underlying energy landscape. Temperatures on the order of 1000 K are often used for sampling the conformations of peptides and proteins, because this temperature is below the temperature at which unwanted cis-trans transitions of the peptide bond frequently occur. In other cases, such as for sampling the conformations of a ligand bound in a protein's active site, much lower temperatures must be used. Otherwise the ligand will dissociate and the simulation will sample the conformations of an unbound ligand rather than those of the bound ligand.

The main advantage of using MD for conformation sampling is that information of molecular forces is used to guide the search process into meaningful regions of the potential. A disadvantage associated with this sampling technique is the fact that high temperature simulations sample not only the regions of interest at room temperature but also regions that are inaccessible to the molecule at room temperature. To overcome this problem the sampled conformations have to be energy-minimized or preferably annealed before being considered as sampled conformations. These methods will be discussed in Section III.

Monte Carlo Simulations

Monte Carlo search methods are stochastic techniques based on the use of random numbers and probability statistics to sample conformational space. The name "Monte Carlo" was originally coined by Metropolis and Ulam during the Manhattan Project of World War II because of the similarity of this simulation technique to games of chance. Today a variety of Monte Carlo (MC) simulation methods are routinely used in diverse fields such as atmospheric studies, nuclear physics, traffic flow, and, of course, biochemistry and biophysics. In this section we focus on the application of the Monte Carlo method for conformational searching.

In performing a Monte Carlo sampling procedure we let the dice decide, again and again, how to proceed with the search process. In general, a Monte Carlo search consists of two steps : (1) generating a new "trial conformation" and (2) deciding whether the new conformation will be accepted or rejected.

Starting from any given conformation we "roll the dice," *i.e.*, we let the computer choose random numbers, to decide what will be the next trial conformation. The precise details of how these moves are constructed vary from one study to another, but most share similar traits. For example, assuming that the search proceeds via polypeptide torsion moves, choosing a new trial conformation could include the following steps. First, roll the dice to randomly select an amino acid position along the polypeptide backbone. Second, randomly select which of the several rotatable bonds in that amino acid will be modified (*e.g.*, the ϕ, ψ or χ torsion angles). Finally, randomly select a new value for this torsion angle from a predefined set of values. In this example it took three separate random selections to generate a new trial conformation. Multiple torsion moves as well as Cartesian coordinate moves are among the many possible variations on this procedure.

Once a new "trial conformation" is created, it is necessary to determine whether this conformation will be accepted or rejected. If rejected, the above procedure will be repeated, randomly creating new trial conformations until one of them is accepted. If accepted, the new conformation becomes the "current" conformation, and the search process continues from it. The trial conformation is usually accepted or rejected according to a temperature-dependent probability of the Metropolis type,

$$p = \begin{cases} e^{\beta \Delta U}, & e^{-\beta \Delta U} < 1 \\ 1, & e^{-\beta \Delta U} \geq 1 \end{cases} \quad \text{or } p = \min[1, e^{-\beta \Delta U}] \qquad \text{...(1)}$$

where $\beta = 1/kT$ and ΔU is the change in the potential energy. This means that if the energy of the new trial conformation is lower than that of the current conformation, $\Delta U < 0$, it is always accepted. But even if the energy of the trial conformation is higher than the current energy, $\Delta U > 0$, there is a certain probability, proportional to the Boltzmann factor, that it will be accepted. To find out whether a higher energy trial conformation is accepted, a random number r in the range [0, 1] is selected and compared to the Metropolis probability defined in Eq. (1). If $r < p$, the conformation is accepted; otherwise it is rejected. This acceptance probability satisfies the principle of detailed balance, ensuring that if the process continues for a long enough time then a stationary solution will be achieved.

In Monte Carlo simulations, just as in MD simulations, temperature plays an important role. In general, MC simulations tend to move toward low energy states. However, at high temperatures (small β values) there is a significant probability of climbing up energy slopes, allowing the search process to cross high energy barriers. This probability becomes significantly smaller at low temperatures, and it vanishes altogether in the limit of $T \rightarrow 0$, where the method becomes equivalent to a minimization process. Thus, high temperature MC is often used to sample broad regions.of conformational space.

As stated above, MC simulations are popular in many diverse fields. Their popularity is due mainly to their ease of use and their good convergence properties. Nonetheless, straightforward and application of MC methods to biomolecules is often problematic due to very low acceptance ratios, which significantly reduce the efficiency of the method. The reason for the low acceptance ratio, *i.e.*, the ratio between accepted MC moves and total MC trial moves, is the compact character of most biomolecules. This means that many move attempts end up "bumping" into other parts of the molecule and are rejected because of the clash. This is true in particular of moves defined in Cartesian coordinates. To partially overcome this problem it is recommended that torsion move-sets be used. Advanced applications of MC methods for conformation sampling often involve various techniques for enhanced sampling.

A special comment must be added regarding the random number generator. Because the MC process is driven by random numbers, it is sensitive to the quality of the random number generator that is being used. The random number generator is supposed to generate uniformly distributed random numbers in the range [0, 1]. But in fact the computer does not generate random numbers at all. It uses a deterministic algorithm to generate a pseudorandom series of numbers with a finite periodicity. A high quality algorithm will have a long enough periodicity that the observed distribution does indeed appear random. However, there are many so-called random number generators on the market that are anything but random. It is good practice to check the random number generator prior to using it in an actual MC simulation. A simple histogram of 10,000 numbers or more will easily show whether or not the resulting distribution is uniform.

Genetic Algorithms

A genetic algorithm (GA) evolves a population of possible solutions through genetic operations, such as mutations and crossovers, to a final population of low energy conformations that minimize the energy function (the fitness function). For the purpose of conformational sampling, the translational, rotational, and internal degrees of freedom are encoded into "genes," which are represented by the real number values of those degrees of freedom. Each individual conformation, named a "chromosome," consists of a collection of genes and is represented by the appropriate string of real numbers. A fitness value (energy) is assigned to each chromosome.

The GA evolutionary process iterates through the following two steps: (1) Generation of a children population from a parent population by means of genetic operators, and (2) a generation update step. The process starts with a random population of initial chromosomes. A new population is generated from the old one by the use of genetic operators. The *mutation* operator changes the value of a randomly selected gene by a random value, depending on the type of gene. The *crossover* operator exchanges a set of genes between two parent chromosomes, creating two children with genes from both parents. Additional genetic operators, such as the *multiple simultaneous mutation* operator or a *migration* operator, which moves individual chromosomes from one subpopulation to another, may also be used. Parents are selected for breeding based on their relative fitness and an external evolutionary "pressure," which directs the process to favor breeding by parents with higher fitness. Following each breeding cycle the chromosome population is updated. A common update scheme is to replace all but the most fit "elite" chromosomes of the original set. For example, a "generation update with an elitism of 2" indicates that all but the two most fit parent chromosomes are replaced. The iteration through the two steps of the evolutionary process continues until convergence or until the maximum number of steps is reached.

Application of GA to conformational searching typically uses the potential energy as the fitness function. The degrees of freedom encoded in the "genes" are usually only a subset of coordinates that are most significant for the conformational search. These may include backbone dihedral angles, side chain dihedral angles, or any other coordinate that may be considered useful. Application of GA to docking problems requires that the three center-of-mass coordinates and the three rotational Euler angles also be encoded in genes.

An advantage of GA for conformational analysis, besides its elegance, is that it is very easy to code and run. In addition, GA usually requires fewer iterations than either MD or MC to generate a large population of low energy conformations. On the other hand, because in every iteration whole populations are propagated, each iteration takes significantly more CPU time than in either of the other two methods. Furthermore, in many complex cases GA is known for its slow convergence.

A successful application of GA to conformation sampling is, for example, as a part of flexible docking. It should be noted, however, that none of the three sampling methods discussed above, MD, MC, and GA, was shown to outperform the other two in any general way. In fact, a comparison of the three methods in the context of flexible docking showed similar efficiency for all three, although specific advantages are likely to exist for particular applications.

Other Search Methods

Many additional search methods have been devised in addition to the basic three just discussed. A few of them are outlined below. Note that whereas some methods, such as parallel tempering and J-walking, are improved or specialized versions of the basic techniques, other methods, such as systematic enumeration and distance geometry, follow different paths altogether.

Systematic enumeration

Systematic sampling is, in principle, the most thorough method for searching molecular conformation al space. The energy is sampled over the entire range of each degree of freedom (typically torsion angles) at regularly spaced intervals. Thus, the sampled conformations lie on an N-dimensional lattice (N being the number of degrees of freedom). Several problems make the systematic sampling procedure irrelevant to most biomolecular systems. A major problem associated with systematic searching is that the number of conformations generated rapidly becomes extremely large even for small molecules. For example, systematic search with six rotatable bonds sampled at 30° increments results in almost 3 million conformations. The number of samples can be reduced by limiting the range of rotation for a symmetrical substituent (*e.g.*, 0-180° for a phenyl group) and/or increasing the rotation step size (all staggered conformations of a saturated $C-C$ bond can be sampled at 120° increments). Another problem arises from the fact that a systematic search is likely to generate a large number of unphysical conformations, in which one pan of the molecule crosses over another part (*e.g.*, a protein backbone that crosses over itself). As a result of the these limitations, systematic search is applied only to small molecules such as small peptides or ligands in the context of flexible docking.

"Pruning" is a sophisticated way to reduce the computational requirements associated with systematic search. Pruning takes advantage of the fact that all conformations along a given branch of the search tree that are downstream from a high energy conformation will also have high energies and need not be calculated. For example, if an irrevocable clash such as the backbone folding over itself occurs after the fifth angle in an eight-angle systematic search, there is no need to continue sampling the three remaining angles. Pruning can be performed using geometrical constraints that must be satisfied for all conformations. An energy cutoff is a simplistic filter that can be applied at the end of the search to reduce the number of conformations stored. With this filter, conformations with energies above the cutoff are discarded. This filter, of course, does not reduce the number of calculations but does help in managing and analyzing the data.

Distance geometry

Distance geometry is a general method for building conformational models of complex molecular systems based on a set of distance constraints. It is a purely geometric technique that generates structures that satisfy the given set of constraints without requiring a starting conformation or an energy function. The distance constraints, which are the input to the method, can be qualitative or approximate. As such they are defined by upper and lower bounds and often into a matrix form. The distance geometry approach converts, or "embeds," the uncertain distance constraints into 3D Cartesian coordinates.

In the basic metric matrix implementation of the distance constraint technique one starts by generating a distance bounds matrix. This is an $N \times N$ square matrix (N the number of atoms) in which the upper bounds occupy the upper diagonal and the lower bounds are placed in the lower diagonal. The matrix is filled by information based on the bond structure, experimental data, or a hypothesis. After smoothing the distance bounds matrix, a new distance matrix is generated by random selection of distances between the bounds. The distance matrix is converted back into a 3D conformation after the distance matrix has been converted into a metric matrix and diagonalized. A new distance matrix randomly generated within the same bounds will result in another possible conformation that satisfies the model, and so forth. The tighter the bounds, the more restricted the search. Naturally, distance geometry is very useful in situations in which many distance constraints are known, in particular for suggesting models that agree with NMR data. This method is also useful for pharmacophore modeling based on known bioactive molecules. However, for general-purpose conformational searching this method is less appropriate unless the search space is limited by known constraints. Computationally, distance geometry is limited to moderately sized systems (up to several thousand atoms) because it requires computationally expensive matrix manipulation.

Parallel tempering and J-walking

Energy barriers that confine the search to limited regions of conformational space restrict both molecular dynamics and Monte Carlo simulations, preventing them from reaching ergodicity. As discussed above, this limitation is often overcome by raising the simulation temperature. The higher temperature allows the simulation to overcome high energy barriers and extend the search space. However, it also causes the simulation to sample regions of conformational space that are irrelevant and inaccessible to room temperature molecules. The methods of parallel tempering and *J*-walking address the problem of how to raise the simulation temperature without wasting computational effort on inaccessible conformations.

The idea behind both parallel tempering and *J*-walking is to incorporate information obtained by ergodic high temperature simulations into low temperature simulations. By periodically passing information from high temperature simulations to low temperature simulations, these methods allow the low temperature system to overcome the barriers between separate regions. In practice, both methods, which can be applied in both MC and MD simulations, require propagation of at least two (and often more) parallel simulations—one at the desired low temperature and another at.a high temperature. For simplicity we shall refer to the Monte Carlo implementation of these techniques; the molecular dynamics implementation is similar in nature.

In *J*-walking the periodic MC trial probability for a simulation at temperature T is taken to be a Boltzmann distribution at a high temperature, T_J ($\beta_J = 1/kT_J$), The jumping temperature, T_J, is sufficiently high that the Metropolois walk can be assumed to be ergodic. This results in the acceptance probability,

$$p = \min[1, e^{-(\beta-\beta_J)\Delta U}] \qquad ...(2)$$

where $\beta = I/kT$, $\beta_J = 1/kT_J$, and ΔU is the change in the potential energy. In practice, at temperature T the trial moves are taken from the Metropolis distribution of Eq. (1) about 90% of the time, with jumps using Eq. (2) attempted about 10% of the time. The jumping conformations are generated with a Metropolis walk at temperature T_J. Several methods have been devised to overcome correlations between the low and high temperature walks.

In parallel tempering, the high and low temperature walkers exchange configurations, unlike in *J*-walking, in which only the high temperature walker feeds conformations to the low temperature walker. By exchanging conformations, parallel tempering satisfies detailed balance, assuming the two walks are long enough. Parallel tempering simulations can be performed for more than two temperatures. It should be noted that the gaps between adjacent temperatures must be chosen so that the exchanges are accepted with sufficient frequency. This means that the distribution characteristics of the two temperatures must overlap so that a conformation generated at one temperature will have a significant probability of being accepted at the other temperature. If the temperature gap is too large, the likelihood of any conformation being accepted by the other simulation is very small.

Conformation Optimization

The structures generated by the various sampling methods often correspond to transient conformations. It is a desirable practice to bring these conformations to nearby local minima, which represent locally stable conformations, before further analysis is performed. The most widely used methods for this purpose are the various minimization techniques. These vary in accuracy and computational efficiency and are useful in many applications as well as conformational analysis.

Minimization methods are used as a tool (often alongside other computational tools) in model building, preparation of the initial structure for molecular dynamics, preparation of structures for normal mode analysis, and more. Because of their importance we go into some detail in discussing the different minimization approaches. Another conformation optimization approach is simulated annealing. In the context of conformational analysis, simulated annealing is often used to optimize conformations generated by high temperature MD or MC simulations. It is also often used as an independent method for minimization and identification of the global minimum in a complex energy landscape.

Minimization

A drop of water that is placed on a hillside will roll down the slope, following the surface curvature, until it ends up in the valley at the bottom of the hill. This is a natural minimization process by which the drop minimizes its potential energy until it reaches a local minimum. Minimization algorithms are the analogous computational procedures that find minima for a given function. Because these procedures are "downhill" methods that are unable to cross energy barriers, they end up in local minima

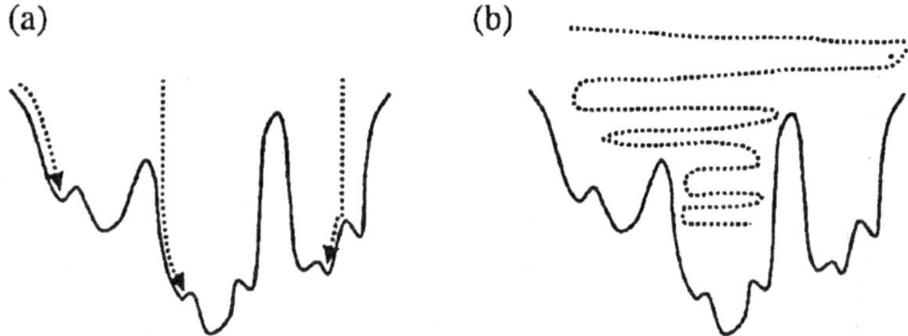

Fig. 5.2. A schematic representation of two optimization schemes. (a) Minimization, which leads to the closest local minimum; (b) simulated annealing, which can overcome intermediate energy barriers along its path.

close to the point from which the minimization process started. It is very rare that a direct minimization method will find the global minimum of the function. This can be easily understood from our water drop analogy. The drop rolls down the slope until it reaches the smallest of hollows and stops there. It will not cross the nearest rise even if the valley on the other side of that rise is much deeper than the hollow it is in now.

In conformational analysis, minimization reduces the potential energy of a given conformation, typically by relieving local strains in the structure. However, the minimized structure will not stray far from the initial conformation. This is why the minimization process is considered an optimization technique and not a search method. Global optimization—namely, reaching the lowest point on the energy surface—cannot be addressed by straightforward minimization.

The goal of all minimization algorithms is to find a local minimum of a given function. They differ in how closely they try to mimic the way a drop of water or a small ball would roll down the slope, following the surface curvature, until it ends up at the "bottom." Consider a Taylor expansion around a minimum point X_0 of the general one-dimensional function $F(X)$, which can be written as

$$F(X) = F(X_0) + (X - X_0)\,F'(X_0) + \frac{1}{2}(X - X_0)^2\,F'')(X_0) + \ldots \quad \ldots(3)$$

where F' and F'' are the first and second derivatives of the function. The extension to a multidimensional function requires replacement of the variable X by the vector X and replacement of the derivatives by the appropriate gradients. Minimization algorithms are classified according to the amount of information about the function that is being used, represented by the highest derivative used by the algorithm. Algorithms that use only the value of the function and no derivative are called "order 0" algorithms. Algorithms that use the first derivative—the slope—of the function are denoted as "order 1" algorithms. Finally, "order 2" algorithms are algorithms that take advantage of the second derivative—the curvature—of the function. As in most computational techniques, higher order methods that use more information about the function are generally more accurate at the price of being computationally more expensive, taking more time and more resources than the lower order methods.

Order 0 algorithms

Order 0 minimization methods do not take the slope or the curvature properties of the energy surface into account. As a result, such methods are crude and can be used only with very simple energy surfaces, *i.e.*, surfaces with a small number of local minima and monotonic behavior away from the minima. These methods are rarely used for macromolecular systems.

Grid searching is an example of an order 0 minimization algorithm. In this method a regular grid (*e.g.*, a cubic grid) is placed over the energy surface and the value of the function at each node is calculated. The grid point with lowest energy is taken to represent the real minimum. The quality of a grid search and the computational effort associated with it depend, of course, on the density of the grid mesh. In some applications the convergence of the method is improved by gradually increasing the density of the grid in the vicinity of the best point of an earlier, coarser grid. This method is inefficient for large molecules and can easily converge to a false minimum. Grid searching as a minimization technique is rarely used in biomolecular studies.

Another order 0 method is the *downhill simplex method.* A *simplex* is a geometrical element consisting, in N dimensions, of $N + 1$ points (vertices) and all their interconnecting line segments, polygonal faces, etc. In two dimensions, a simplex is a triangle; in three dimensions it is a tetrahedron. The downhill simplex method starts with $N + 1$ points, defining an initial simplex. The energy of each point is evaluated, and the method moves the point of the simplex from where the function is largest through the opposite face of the simplex to a lower point. Usually such "reflections" conserve the volume of the simplex, but when it can

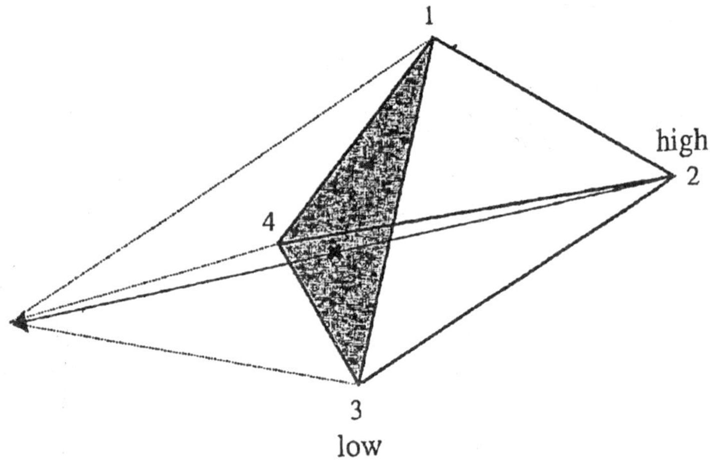

Fig. 5.3. A representative step in the downhill simplex method. The original simplex, a tetrahedron in this case, is drawn with solid line. The point with highest energy is reflected through the opposite triangular plane (shaded) to form a new simplex.

do so it will expand or contract the simplex to minimize the function. The minimization proceeds by taking the highest point in the new simplex and reflecting it through the opposite face and so forth. The simplex method is sometimes used for crude placement of starting conformations, for example in the context of docking.

Order 1 algorithms

Order 1 algorithms, which represent a fair balance between accuracy and efficiency, are the most commonly used minimization methods in macromolecular simulations. As indicated by their name, these algorithms use the gradient of the function to direct the minimization process toward the nearest local minimum. They use information about the *slope* of the function but do not include information about its *curvature*. Thus, to compensate for the deficiency of curvature data, all order 1 minimization methods employ a stepwise iterative scheme. The iterations are used for recalculating the gradient and correcting the minimization approach pattern following changes in the direction of the slope.

In general, order 1 methods iterate over the following equation in order to perform the minimization until it converges or until it reaches a preset maximum number of steps :

$$\hat{r}_k = \hat{r}_{k-1} + \lambda_k \hat{S}_k \qquad ...(12.4)$$

where r_k is the new position at step k, r_{k-1} is the position at the previous step $k - 1$, λ_k is the size of the step to be taken at step k, and S_k is the direction of that step. Various methods differ in the way they choose the step size and the step direction, *i.e.*, in the way they try to compensate for the lack of curvature information. For example, the *steepest descent* method uses a variable step size for this purpose, whereas the *conjugated gradients* method takes advantage of its memory of previous steps. These two popular algorithms are described next.

Steepest descent

Steepest descent (SD), which is the simplest minimization algorithm, follows the direction of the instantaneous gradients calculated at each iteration. This means that once the gradient of the energy function, g_k, at the current position is calculated, the minimization step is taken in the direction opposite to it (*i.e.*, in the direction of the force),

$$\hat{S}_k = -\hat{g}_k = -\nabla U(r) \qquad ...(5)$$

The step size, λ_k, is adjusted at each iteration to compensate for the lack of curvature information. If the energy of the new conformation is lower than that of the previous one, the algorithm assumes that the approach direction is correct and increases the step size Xj. by a small factor (often by the factor 1.2) to improve efficiency. However, if the energy of the new conformation turns out to higher, indicating that the real path has curved away from the current minimization direction, the step size λ_k is decreased (often by the factor 0.5) to allow the algorithm to correct the direction more efficiently. Because finite step sizes are used, the method does not flow smoothly down to the minimum but rather jitters around it. Furthermore, SD's imprecise approach to the minimum usually means that the method does not converge to one point and gets stuck in a limit cycle around the minimum. This means that SD often gets close to the minimum but rarely reaches it. Despite the relatively poor convergence of SD it is an efficient minimization procedure that is very useful for crude local optimization, such as relieving bad contacts in an initial structure before a dynamics simulation or as the first phase of a more complex minimization scheme.

Conjugated gradients

To compensate for the missing curvature information, the conjugated gradients (CG) method makes use of its "memory" of gradients calculated in previous steps. The first step is taken in the direction of the force,

$$\hat{S}_1 = -\hat{g}_1 \quad ...(6)$$

but for all subsequent iterations, $k > 1$, the direction of the minimization step is determined as a weighted average of the current gradient and the direction taken in the previous iteration, *i.e.*,

$$\vec{S}_k = -\vec{g}_k + b_k \vec{S}_{k-1} \quad ...(7)$$

The weight factor b_k is calculated as a ratio of the squares of the current and previous gradients.

$$b_k = |g_k|^2/|g_{k-1}|^2 \quad ...(8)$$

For quadratic surfaces of dimension n, it can be shown that the method of conjugated gradients is very efficient, converging on the nth step. Nonetheless, even for nonquadratic surfaces such as molecular energy surfaces, the CG method converges much better than SD. A numerical disadvantage of the gradient memory employed in CG is that it accumulates numerical errors along the way. This problem is usually overcome by restarting the minimization process every so often; that is, at given intervals the gradient memory is wiped out by setting b_k equal to zero.

Order 2 algorithms

Order 2 minimization algorithms, which use the second derivative (curvature) as well as the first derivative (slope) of the potential function, exhibit in many cases improved rate of convergence. For a molecule of N atoms these methods require calculating the $3N \times 3N$ Hessian matrix of second derivatives (for the coordinate set at step k)

$$[H_k]_{ij} = \frac{\partial^2 U(r)}{\partial r_i \, \partial r_j} \quad ...(9)$$

in addition to the $3N$ vector of first derivatives g_k discussed above.

Newton-Raphson

The Newton-Raphson minimization method is an order 2 method based on the assumption that, near the minimum, the energy can be approximated by a quadratic function. For a one-dimensional case, assuming that $F(X) = a + bX + cX^2$, the Newton-Raphson minimization can be formulated as

$$X^* = X - F'(X)/F''(X) \qquad \text{...(10)}$$

where X^* is the minimum, X is the current position, and F' and F'' are the first and second derivatives at the current position. Namely, for a quadratic function, this algorithm finds the minimum in a single step. This conclusion can be generalized to the multidimensional case :

$$\vec{S}_k = -H_k^{-1}\,\vec{g}_k \qquad \text{...(11)}$$

For nonquadratic but monotonic surfaces, the Newton-Raphson minimization method can be applied near a minimum in an iterative way.

There are several reasons that Newton-Raphson minimization is rarely used in mac-romolecular studies. First, the highly nonquadratic macromolecular energy surface, which is characterized by a multitude of local minima, is unsuitable for the Newton-Raphson method. In such cases it is inefficient, at times even pathological, in behavior. It is, however, sometimes used to complete the minimization of a structure that was already minimized by another method. In such cases it is assumed that the starting point is close enough to the real minimum to justify the quadratic approximation. Second, the need to recalculate the Hessian matrix at every iteration makes this algorithm computationally expensive. Third, it is necessary to invert the second derivative matrix at every step, a difficult task for large systems.

Adopted basis Newton-Raphson

A derivative method that is suited for large systems such as proteins is the *adopted basis Newton-Raphson* (ABNR) algorithm. Instead of calculating the full multidimensional curvature (*i.e.*, the full Hessian matrix) at each minimization step, the ABNR method limits its calculation to a small subspace of dimension s in which the system has made the most progress in past moves. The idea is to add curvature information only in those directions where it is likely to contribute most. This way the system moves in the best direction in the restricted subspace. Because the dimensionality s of the subspace is taken to be between 4 and 10, ABNR is an efficient minimization method comparable in CPU time to the order 1 methods such as the conjugated gradient.

ABNR is also comparable to CG in terms of convergence. Since the method is not self-starting, the first steps are taken with an order 1 method, usually the steepest descent method.

Minimization protocol

The foregoing discussion highlights the fact that the different minimization algorithms have relative strengths and weaknesses. As an example, Figure 5.4 compares the results of minimizing the same protein with the steepest descent (SD) and conjugated gradients (CG) methods. It is evident that although initially SD reduces the energy faster than CG, in the long run the latter outperforms the former. A similar result is obtained when CG is replaced by ABNR.

To optimize the minimization procedure it is usually best to combine several algorithms into a single minimization protocol, taking advantage of their relative strengths. A good minimization scheme will usually start with SD and then use CG or ANBR to finish the job. The number of steps to be used in each phase depends on the goal of the minimization and on the character of the system. When high quality minimization is required, the minimization can be completed with Newton-Raphson. The termination criterion is usually defined in terms of the gradient

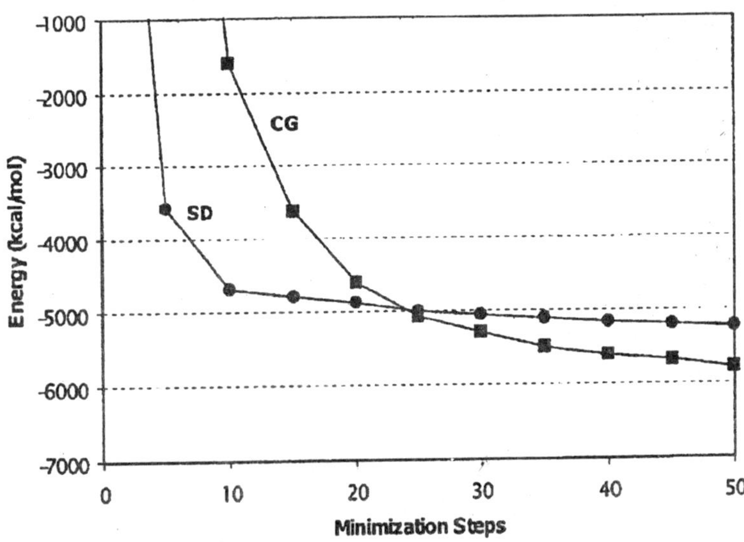

Fig. 5.4. A comparison of steepest descent (SD) minimization and conjugated gradient (CG) minimization of the same protein.

RMS (GRMS), which is defined as the root-mean-square of all 3*N* gradients (or forces).

Simulated Annealing

Simulated annealing is a popular method that is often used for global optimization, i.e., finding the global minimum of a potential energy surface. The method takes its name from the natural annealing process in which a glass or a metal is first heated and then slowly cooled into a stable low energy state. The key factor in this process is slow cooling. If the cooling is done too fast, the materials will end up in unstable brittle states. Alternalively phrased, heating up the system shakes and rattles the molecule around the energy landscape, infusing it with thermal energy, analogous to kT, enabling it to jump out of its initial local minimum. The gradual cooling that is subsequently applied decreases the amplitude of these shakes, bringing the details of the energy surface back into focus and causing the system to slowly settle down to a lower energy minimum. In fact, we see that simulated annealing bridges between the high temperature conformational sampling simulations that are insensitive to the energy barriers and the low temperature situations, which are sensitive to the details of the energy landscape.

Simulated annealing can be easily implemented in both molecular dynamics and Monte Carlo simulations. In molecular dynamics, the temperature is controlled through coupling to a heat bath; with simulated annealing, the temperature of the bath is decreasing gradually. In Monte Carlo the trial move is accepted or rejected according to a temperature-dependent probability of the Metropolis type. In simulated annealing MC, the temperature used in the acceptance probability is gradually decreased. It should be noted that it is not necessary to anneal all the way to 0 *K*, because once the kinetic energy kT gets below the characteristic barrier height, a significant change cannot occur. Thus, many simulated annealing protocols cool to room temperature (or somewhat below) and are followed by a local minimization algorithm to remove the excess energy. Specific implementations vary in cooling schedules, initial temperatures, the possibility of repeated heating "spikes," etc.

Although simulated annealing is often considered a global optimization method, this is not the case when biomolecules are concerned. It can be shown that in systems characterized by a broad distribution of energy scales, a simulated annealing trajectory (either MD or MC) will have to be extremely long before it is able to find the global minimum. Since the energy landscape of proteins is broadly distributed and rough, this means that simulated annealing is an inefficient and infeasible strategy for protein folding. Nonetheless, even with biomolecules, simulated annealing remains a very useful method for local optimization. Its advantage is that, unlike direct minimization, which takes the molecule only as far as the nearest local minimum, simulated annealing is able to locate lower local minima further away from the initial conformation. In the context of conformational analysis, simulated annealing is often used in conjunction with a high temperature sampling simulation. Each of the molecular structures generated by the high temperature simulation is first annealed back to room temperature before it is included in the conformational sample and subjected to further analysis.

Conformational Analysis

To extract the conformational properties of the molecule that is being studied, the conformational ensemble that was sampled and optimized must be analyzed. The analysis may focus on global properties, attempting to characterize features such as overall flexibility or to identify common trends in the conformation set. Alternatively, it may be used to identify a smaller subset of characteristic low energy conformations, which may be used to direct future drug development efforts. It should be stressed that the different conformational analysis tools can be applied to any collection of molecular conformations. These may be generated by the above sampling techniques but can also have an experimental origin, such as NMR models or different X-ray structures of the same molecule (or analogous molecules).

Similarity Measures

A similarity measure is required for quantitative comparison of one structure with another, and as such it must be defined before the analysis can commence. Structural similarity is often measured by a root-mean-square distance (RMSD) between two conformations. In Cartesian coordinates the RMS distance d_{ij} between conformation i and conformation j of a given molecule is defined as the minimum of the functional

$$d_{ij} = \left[\frac{1}{N} \sum_{k=1}^{N} |r_k^{(i)} - r_k^{(j)}|^2 \right]^{1/2} \qquad ...(12)$$

where N is the number of atoms in the summation, k is an index over these atoms, and $r_k^{(i)}, r_k^{(j)}$ are the Cartesian coordinates of atom k in conformations i and j. The minimum value of Eq. (12) is obtained by an optimal superposition of the two structures. The resulting RMS distances are usually compiled into a distance matrix Δ, where the elements Δ_{ij} are the RMS distances between conformations i and j.

Since the summation in Eq. (12) may be on any subset of atoms, it can be fine-tuned to best suit the problem at hand. The summation may be over the whole molecule, but it is very common to calculate conformational distances based only on non-hydrogen "heavy" atoms or, in the case of proteins, even based on only the backbone C_α atoms. Alternatively, in a study related to drug design one may consider, for example, focusing only on atoms that make up the pharmacophore region or that are otherwise known to be functionally important.

The conformational distance does not have to be defined in Cartesian coordinates. For comparing polypeptide chains it is likely that similarity in dihedral angle space is more important than similarity in Cartesian space. Two conformations of a linear molecule separated by a single low barrier dihedral torsion in the middle of the molecule would still be considered similar on the basis of dihedral space distance but will probably be considered very different on the basis of their distance in Cartesian space. The RMS distance is dihedral angle space differs from Eq. (12) because it has to take into account the 2π periodicity of the torsion angle,

$$d_{ij} = \left\{ \frac{1}{N} \sum_{k=1}^{N} \min[(\theta_k^{(j)} - \theta_k^{(j)})^2, (2\pi - \theta_k^{(j)} + \theta_k^{(j)})^2 \right\}^{1/2} \qquad ...(13)$$

where N is the number of dihedral angles in the summation and $\theta_k^{(j)}, \theta_k^{(ij)}$ are the values of the dihedral angle θ_k in the two structures. As with the Cartesian distance, any appropriate subset of dihedral angles may be used, ranging from only the backbone ϕ, ψ, angles to a full set that includes all the side chain χ angles.

It is up to the researcher to decide whether to use a Cartesian similarity measure or a dihedral measure and what elements to include in the summation. It should be stressed that while the RMS distances perform well and are often used, there are no restrictions against other similarity measures. For example, similarity measures that emphasize chemical interactions, hydrophobicity, or the relative orientation of large molecular domains rather than local geometry may serve well if appropriately used.

Cluster Analysis

The distance matrix Δ, which holds the relative distances (by whatever similarity measure) between the individual conformations, is rarely informative by itself. For example, when sampling along a molecular dynamics trajectory, the Δ matrix can have a block diagonal form, indicating that the trajectory has moved from one conformational basin to another. Nonetheless, even in this case, the matrix in itself does not give reliable information about the size and shape of the respective basins. In general, the distance matrix requires further processing.

Cluster analysis is a common analytical technique used to group conformations. This approach highlights structural similarity, as defined by the distance measure being used, within a conformational sample. Starting from one selected conformation (often that of the lowest energy), all conformations that are within a given cutoff distance from this structure are grouped together into the first cluster C_1. Next, one of the conformations that were not grouped into the first cluster is selected, and a new cluster is formed around it.

The process continues until all the conformations in the sample are assigned to a cluster C_i. This process often generates overlapping clusters, that is, clusters with nonzero intersection $C_i \cap C_j \neq 0$, The overlapping clusters are typically treated in one of two ways: (1) Group together the overlapping clusters C_i and C_j to form a single large cluster that is their union $C_i, \cup C_j$ or (2) make the overlapping clusters disjoint by removing their intersection $C_i \cap C_j$ from one of the clusters, typically the one that started with a higher energy conformation. Since the optimal cutoff distance by which to cluster the conformations is not a priori known, cluster analysis is usually performed hierarchically. Starting with a short cutoff the analysis is repeated

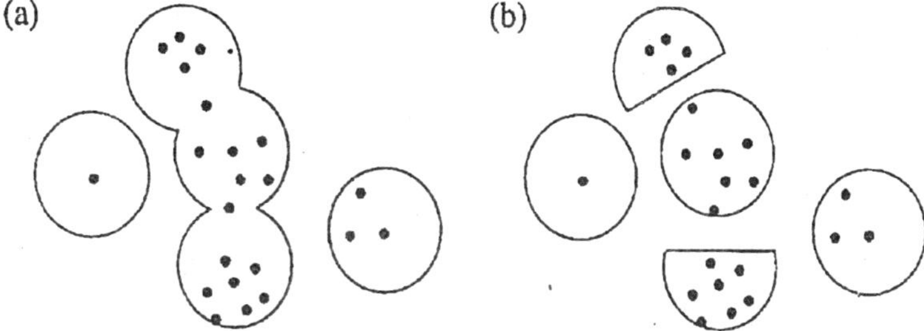

Fig. 5.5. A schematic representation of two clustering methods, in which each point represents a single molecular conformation and the circles are the similarity cutoff distances used to define the clusters. (a) Three clusters are defined when overlapping clusters are grouped together. (b) Five clusters are defined when the overlaps are removed from one of the overlapping clusters.

again and again, each time with a larger cutoff distance. The results are often represented as a dendogram.

In many conformational studies, cluster analysis is used as a way to focus future effort on a small set of characteristic conformations. One conformation, typically the lowest energy one, is picked from each of the highly populated conformational clusters. The resulting small number of distinctly different conformations are then used as starting points for further computational analysis (such as free energy simulations) or as a basis for generating a pharmacological hypothesis used for directing future drug development. It should be noted, however, that conformational clusters generated by the above procedure do not necessarily represent the correct basin structure of the underlying energy landscape.

It has been shown that similar conformations that belong to adjacent energy basins separated by high energy barriers are incorrectly grouped together by the straightforward cluster analysis.

Principal Component Analysis

An inherent problem associated with conformational analysis is the high dimensionality of the molecular conformational spaces. An *N*-atom molecule has 3*N* degrees of freedom, and its corresponding conformational space is (3*N* – 6)-dimensional. As a result, even relatively small molecules have very large conformational spaces, making them difficult to analyze. For example, a small heptapeptide may have a 100-dimensional or even 150-dimensional conformational space, depending on its precise amino acid composition. Principal component analysis (PCA) is a computational tool that reduces the effective dimensionality of molecular conformational spaces while retaining an accurate representation of the interconformational distances. This task is accomplished by projecting the original multidimensional data onto an optimal low-dimensional subspace, allowing visual inspection of conformational spaces and of dynamic trajectories that traverse these spaces. Principal component analysis (PCA) was introduced to protein simulations under the name quasi-harmonic analysis by Ichiye and Karplus and is becoming widely used for a variety of applications involving sampling and visualization of conformational spaces.

How does principal component analysis work? Consider, for example, the two-dimensional distribution of points shown in Figure 5.6. This distribution clearly has a strong linear component and is closer to a one-dimensional distribution than to a full two-dimensional distribution. However, from the one-dimensional projections of this distribution on the two orthogonal axes X and Y you would not know that. In fact, you would probably conclude, based only on these projections, that the data points are homogeneously distributed in two dimensions. A simple axes rotation is all it takes to reveal that the data points are preferentially spread along one dimension, as reflected by the broad distribution along one of the new axes and a narrow

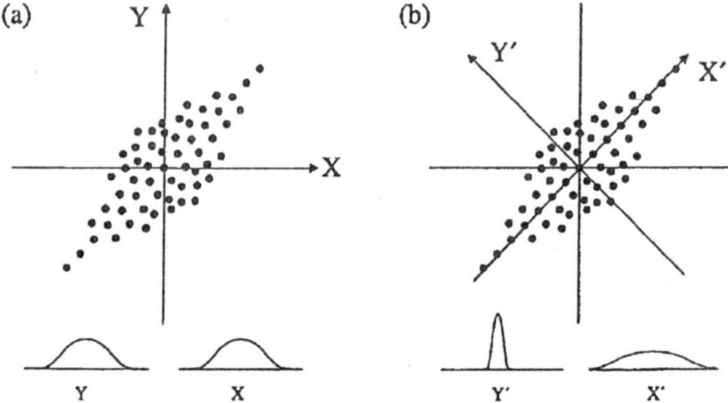

Fig. 5.6. (a) A two-dimensional distribution of points and their one-dimensional projections on the original axes. (b) The same distribution of points and their 1D projections on the new axes set obtained by PCA by using a similarity transformation.

distribution along the other. The above procedure is what is done by PCA. Starting from a multidimensiorral distribution of data points (in our case, of molecular conformations), PCA performs a similarity transformation on the original axes to find a new set of axes that best fits the data. The first new axis is selected such that the variance of the distribution along it is the largest possible. The second axis is placed orthogonal to the first in the direction of the second largest variance of the distribution, and so forth. In this new axes set it is usually possible to identify a low-dimensional subspace that captures most of the relative distances between individual conformations.

In general, two related techniques may be used: principal component analysis (PCA) and principal coordinate analysis (PCoorA). Both methods start from the $n \times m$ data matrix M, which holds the m coordinates defining n conformations in an m-dimensional space. That is, each matrix element M_{ij} is equal to q_{ij}, the jth coordinate of the ith conformation. From this starting point PCA and PCoorA follow different routes.

Principal component analysis (PCA) takes the m-coordinate vectors q associated with the conformation sample and calculates the square $m \times m$ M^TM matrix, reflecting the relationships between the *coordinates*. This matrix, also known as the covariance matrix C, is defined as

$$C = \langle (q - \langle q \rangle)\,(q - \langle q \rangle)^T \rangle \qquad ...(14)$$

where the averaging is over the conformation sample (in Cartesian space $m = 3N$ for an N-atom molecule). The covariance matrix C is diagonalized to obtain the eigenvectors that capture most of the variation in atomic position fluctuations.

Principal coordinate analysis (PCoorA), on the other hand, operates on the square $n \times n$ MM^T matrix, reflecting the relationships between the *conformations*. The elements of this matrix, also known as the distance matrix Δ, are distances d_{ij} between two conformations i andy j [such as those defined in Eqs. (12) and (13). Since the distances d_{ij} can also be obtained from the $n \times n$ matrix A of latent roots (eigenvectors), one can use this matrix for the projection, defining $A_{ij} = -1/2\, d_{ij}^2$ and $A_{ii} = 0$ (for $i, j = 1, 2, ..., n$). To guarantee that the matrix A has a zero root (and thus guarantee that it corresponds to a real configuration) it is "centered," so that the sum of every row and the sum of every column of A is zero. This centering, which does not alter the distances d_{ij}, is defined as

$$A^*_{ij} = A_{ij} - \langle\langle A_{ij}\rangle_i - \langle A_{ij}\rangle_j + 2\langle A_{ij}\rangle_{ij} \qquad \text{...(15)}$$

where $\langle\cdot\rangle_k$ is the mean over all specific indices $k = i, j, ij$. The centered matrix $A*$ is diagonalized using standard matrix algebra to obtain the latent eigenvectors and the diagonal matrix of eigenvalues. The resulting eigenvalues (normalized) give the percentage of the projection of the original distribution on the new coordinate set, and the eigenvectors (scaled by their corresponding eigenvalues) give the new coordinates of the original points in the new axes set.

It should be stressed that PCA and PCoorA are dual methods that give the same analytical results. Using one or the other is simply a matter of convenience, whether one prefers to work with the covariance matrix C or with the distance matrix Δ.

As stated earlier, the main motivation for using either PCA or PCA is to construct a low-dimensional representation of the original high-dimensional data. The notion behind this approach is that the effective dimensionality of a molecular conformational space is significantly smaller than its full dimensionality (3N-6 degrees of freedom for an N-atom molecule). Following the PCA procedure, each new axis k is associated with a normalized eigenvalue λ_k that reflects the relative weight of that axis in reproducing the original data. An axis with a high λ_k value is significant for the projection, whereas axes with small λ_k values are insignificant. By sorting the new axes according to their λ_k weight it is possible to select a small subset of effective coordinates that capture most of the conformational relationships of the original high-dimensional space. The quality of such a projection can be estimated by the average difference between conformation distances reconstructed in the low s-dimensional subspace $d_{ij}^{(s)}$ and the original distance d_{ij}. The reconstructed distances are defined as

$$d_{ij}^{(s)^2} = \sum_{k=1}^{s} (Q_{ik} - Q_{jk})^2 \qquad \text{...(16)}$$

where Q_{ik} is the coordinate of the ith conformation along the kth new (principal) axis. It can be shown that the average deviation of the distances in s dimensions from the exact distances is given by the sum of the first s eigenvalues,

$$\langle d_{ij}^2 - d_{ij}^{(s)^2} \rangle_{ij} = \sum_{k=1}^{s} \lambda_k \qquad \text{...(17)}$$

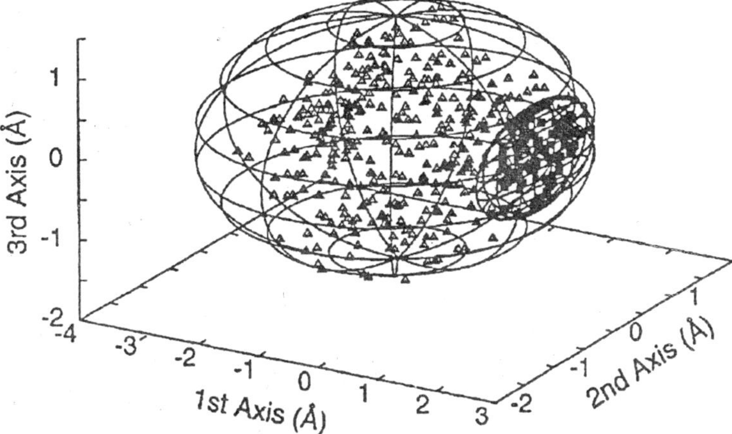

Fig. 5.7. A joint principal coordinate projection of the occupied regions in the conformational spaces of linear $(Ala)_6$ (triangles) and its conformational constraint counterpart, cyclic-$(Ala)_6$ (squares), onto the optimal 3D principal axex.

Thus by summing the normalized eigenvalues of the first s dimensions one can judge the quality of a projection onto that subspace.

Fortunately, it was found that in polypeptide systems the effective dimensionality of conformational spaces is significantly smaller than the dimensionality of the full space, with only a few principal axes contributing to the projection. In fact, in many cases a projection quality of 70-90% can be achieved in as few as three dimensions, opening the way for real 3D visualization of molecular conformational space. Figure 5.7 shows a 3D visualization of the conformational spaces of two hexapeptides, $(Ala)_6$ and cyclic-$(Ala)_6$, jointly projected on the same principal coordinate set. The comparison shows that the conformation volume occupied by the linear peptide is about 10 times larger than the conformation volume occupied by its conformationally constrained counterpart. Quantifying relative flexibility of analogous peptides through joint principal projections was shown to be useful, for example, in predicting their relative bioactivity.

6

DATA MANAGEMENT

Bioinformatics methods aim to extract information about biological functions and mechanisms from computer-held biological data. These data range from genomic sequence data to the transcriptomics and proteomics data discussed in previous chapters. Data management techniques enable these increasingly huge amounts of data to be stored and extracted from computers reliably and efficiently. Most of these techniques were developed for business applications, where the data are structured and it is clear what object any particular piece of data relates to, such as in the case of recording a customer's name, address and account balance. However, managing biological data and extracting useful information from them poses particular problems for the following reasons.

1. Biological data come in many different forms: structured, as in GenBank or PDB file formats; unstructured, as in biomedical literature abstracts; sequence, as in genomic or protein sequences; or more complex, as in biochemical pathway data.
2. There is no single biological object to which the data refer: individual users of the data have their own perspective depending on the science they are carrying out, ranging from the atomic level to that of whole organisms.
3. Similarly, the biological functions and mechanisms to which data refer are also dependent on the perspective of the user of the data.

4. Even if the appropriate biological object can be identified, there are many incompatible formats and nomenclatures for the data about that object and its functions.
5. The data about any given biological object are likely to be held at many different sites,each with a different approach to coping with the previous problems.

Data Management Concepts

Databases and Database Software

The term *database* can be used simply to mean any collection of data stored in a computer, perhaps even in a single file. Biological data have indeed frequently been held in simple files often with tags to indicate the meaning of a particular piece of data as, for example, in the case of a GenBank file. Files may also have a fixed record format, as in the case of a PDB file. As discussed below, problems often arise with file-based databases, and this has led to the development of database management techniques to overcome such problems.

Using these techniques, a database is still a collection of data stored in a computer, but the storage of and access to the data is controlled by specialized software called a *database management system* (DBMS). In practice, even biological data which have traditionally been in a file format, such as GenBank or PDB data, are in reality now managed by a DBMS and the familiar file formats are retained as one way of making such data available to scientists.

Why are DBMSs Useful?

Scientists have traditionally written programs in languages such as Perl, C or C++ to extract data relevant to their work from files, perform processing to generate new data of interest, and then write those data to a new file. This way of working leads to the following problems.

1. Data are not shared. The data generated by each scientist are stored in different files, the format of which will be dependent on the precise programs written by the scientist. It will not be possible for other scientists to share those data without knowing of the existence of the files and their format.
2. Data are not integrated. The new files generated by each scientist have no relationship to the files of data from which they were derived, nor with the many files being generated

by other scientists. The information which would be available if the data stored in the separate files were integrated and inter-related is lost.

3. Data are duplicated. The same or closely related data are generated many times by different programs, and stored in different files. This inevitably leads to problems when revisions are made to one file of generated data, perhaps to correct an error, but the other files of identical or closely related data are not revised. Multiple versions of data which should be consistent are then inconsistent.
4. Programs are dependent on file formats. If the format of a file is changed, perhaps to add some new type of data, then the programs which access that file have to be modified to take account of the new format.

DBMSs were developed to overcome these problems. They do this in the following ways.

1. Data are held in a single shared resource: the database. Each scientist still writes programs to access data, but the format of the data is defined centrally for the database rather than being determined by an individual scientist's program. Further, if the data generated by the scientist are themselves stored in the database, they can be shared by other scientists.
2. Data are integrated in the database. When data are stored in a DBMS-managed database, the DBMS enables the relationships between all the data to be explicitly represented, and the information inherent in those relationships is available.
3. Data need not be duplicated. In reality data *are* almost always duplicated in a DBMS-managed database, but the duplication is controlled by the DBMS. The reasons for this duplication are to enhance performance, and to enable recovery of lost data if some failure results in their loss.
4. Programs are independent of file formats. The format of the data may be changed or new types of data added to a DBMS-managed database, but existing programs which have no requirement to access the new types of data need not be modified: the DBMS itself ensures existing programs still work as though there had been no change to the data formats.

The difference between file-based and DBMS-managed data is illustrated in Figure 6.1. The upper half of the figure shows file-based 'databases': many different programs access data in

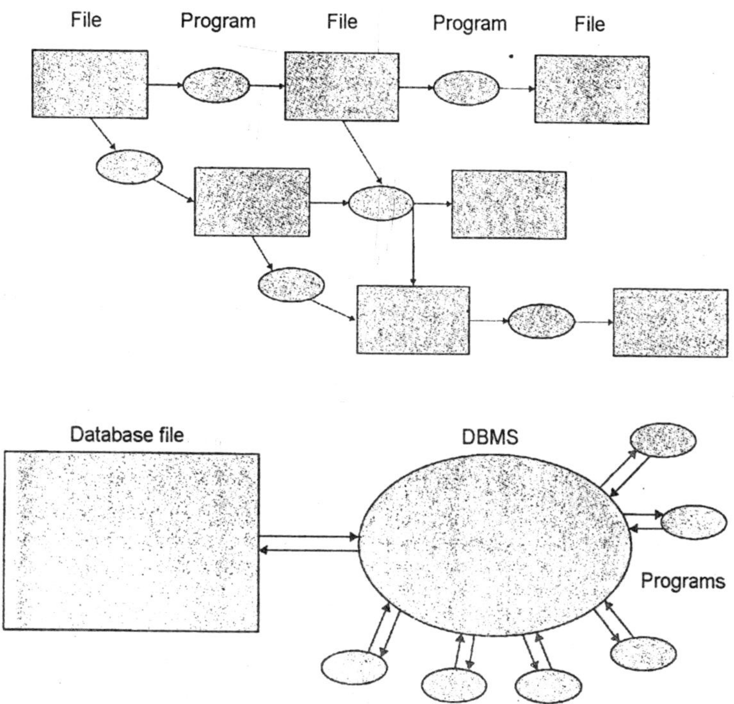

Fig. 6.1. File-based v DBMS-based data.

multiple files and write out new data to yet more files. It becomes an impossible task to maintain data as a shared, integrated, consistent resource. The lower half of the figure shows DBMS-managed data. Programs now access data in a shared, integrated database by using the services of a DBMS, which is able to ensure data is shared, integrated and consistent.

It might be imagined from the above that having data managed by a DBMS rather than using ordinary files is always advantageous. Indeed, many scientists use a DBMS as though the very fact of having data in a 'real' database gives benefits. In fact there are disadvantages in using a DBMS including the following.

1. A DBMS is expensive. This can include the cost of purchasing the DBMS software, but equally importantly a DBMS is a large piece of software which requires more computer resources such as memory than conventional programs accessing files. File-based access to data does not incur these costs.

2. DBMSs are complex. If a scientist writing a Perl program to access a file finds the program is not working as expected, there is every chance that the source of the problem can be discovered and corrected. If the same program is using a DBMS, there are many areas of the DBMS's operation which are complex and can only be corrected with specialist knowledge of the DBMS.
3. Access to data by a DBMS may be slow. The DBMS software is potentially carrying out a great deal of work to enable access to data reliably by many different programs concurrently and there is an inevitable overhead associated with this. A skilled programmer will always be able to write a program accessing file-based data more quickly.

In summary, the advantages of a DBMS are that it provides shared, integrated access to data, with the ability to modify the types and formats of stored data for the needs of new applications, while allowing existing applications to continue working without modification. Usually these are significant advantages to a scientist with data, but if the costs arising cannot be justified, then it makes sense to continue working with file-based data.

Relational and other Databases

Another problem with file-based data is that it can be difficult to understand what information is contained in the file. A PDB file, for example, certainly contains much information about the 3D structure of a macromolecule, but it is no simple matter to understand from the file what are the biological and other entities and their relationships inherent in the data. An important feature of a DBMS is that it supports a *data model* which makes it easier to understand what information the database contains. Many different types of data model have been developed since the 1960s, but for over 10 years the *relational model* has been dominant, and DBMSs supporting the relational model are currently the most widely found.

The essential characteristic of the relational model is that all data are represented as being stored in tables. Some columns from two tables, PDB_DOM and PDB_DOM_SEG are shown with some example rows in each. PDB_DOM records the CATH structural classification for each identified domain in the PDB. A domain may encompass non-contiguous sections of a chain, each

PDB_DOM

PDBCODE	CHAIN	DOMAIN	C	A	T	H	S35	S95	S100
1lga	A	1	1	10	520	10	1	2	1
1lga	A	2	1	10	420	10	2	2	1
1lga	B	1	1	10	520	10	1	2	2
1lga	B	2	1	10	420	10	2	2	2

PDB_DOM_SEG

PDEBCODE	CHAIN	DOMAIN	SEGMENT	RESNUM1	RESNUM2
1lga	A	1	1	11	144
1lga	A	1	2	269	299
1lga	A	2	1	145	268
1lga	A	2	2	300	339
1lga	B	1	1	11	144
1lga	B	1	2	269	296
1lga	B	2	1	145	268
1lga	B	2	2	297	339

Fig. 6.2. Example relational table data.

such section being known as a domain segment. PDB_DOM_SEG records the start and finish residue numbers of each domain segment. Once the decision has been made to use a relational database, there are a large number of different relational DBMSs which are available. These range from relatively simple products such as *MySQL* through products with more features such as *PostgreSQL*, both of which may be obtained freely over the World Wide Web, to more complex products such as *Oracle* which are intended for commercial use and can be expensive to buy. These products all have the essential characteristic of a relational DBMS in that the representation of data is in the form of tables; where they differ is in the sophistication of their features to enable the management of large amounts of data accessed concurrently by many users securely, reliably and efficiently.

While a financial institution would need the levels of security and reliability supported by a product such as Oracle, for many scientists the features supported by free software such as MySQL and PostgreSQL are entirely adequate for their needs. Indeed, the simpler products, without the overhead of features found in more complex products, may have performance advantages for straightforward access to data. Further reading about DBMS products can be found in the books by Yarger *et al.* (MySQL), Worsley and Drake (PostgreSQL) and Abbey *et al.* (Oracle). The relational model has the great advantage of being a very simple way of representing data, but when there are complex relationships between entities, as is the case for many biological data, it can be difficult to understand how the tables represent those relationships. Each PDB structure may have many domains, and each PDB structure may have many chains.

The relationship between domains and chains is a many-many one: namely, each domain may be related to many chains, while each chain may be related to many domains. However, the nature of these relationships is not clearly represented in the tables shown. It is desirable that any data model makes it easy to understand what relationships are represented therein, and the relational model does not always achieve this. An alternative data model, the *object-oriented model,* is supported by some DBMSs. This represents objects and the relationships between objects directly, as well as controlling the way in which objects may be created and accessed in the database. So, for example, rather than having data about a protein domain spread across many tables as would be the case in a relational database, in an object-oriented database all the information about a protein domain, including what a domain is related to and the ways in which domain data may be created and accessed, would be brought together in the definition of a protein domain object. Some DBMSs have been designed specifically for managing scientific data. One having features specifically for managing genetic map, physical map and sequence data is *AceDB,* originally designed for the C. *elegans* sequencing project.

AceDB uses an object-oriented model with each object having a tree-like hierarchical structure. Graphical interfaces for displaying the map and sequence data come with the DBMS. Although an object-oriented model can be well-suited to representing data with complex relationships, the simplicity of the relational model and the strengths of well-established relational DBMSs in supporting reliable, shared access to integrated data have ensured their continued wide use.

Data Management Techniques

Whatever relational DBMS product is used, the fundamental requirement of any user is knowing how to access the data in the database. Those scientists developing their own data bases additionally need to know how to design and create a relational database. Both of these tasks require an understanding of how a relational DBMS works, in order that the database may be used in an efficient way. The techniques underpinning such use of a relational database are now examined.

Accessing a Database

The standard language for accessing a relational database is *SQL* (Structured Query *Language*): no matter which relational DBMS product is used, SQL will be supported. In fact SQL is more than just a language to access the data in a relational database, it is also the language used to create the tables in a database and insert, modify or delete rows of data in those tables. SQL has a SELECT statement to retrieve rows from tables. For example

```
SELECT C, A, T, H
FROM PDB_DOM
WHERE PDBCODE = `1lga'
AND CHAIN = `B'
```

This query has three parts or clauses, a SELECT clause, a FROM clause and a WHERE clause. Every SELECT statement has SELECT and FROM clauses, while the WHERE clause is optional. If this SELECT statement were executed with the example tables, the rows retrieved forming the result would be:

C	A	T	H
1	10	520	10
1	10	420	10

The SELECT clause specifies which values in particular appear in each row in the query result - in this case C, A, T and H values. The FROM clause of the query specifies that data are to be retrieved by accessing the PDB_DOM table. If the query had no WHERE clause, the C, A, T and H values from every row in PDB_DOM would appear in the result. If the query does have a WHERE clause, as in this example, only rows which satisfy the WHERE clause conditions produce values in the query result.

Normally queries will require access to data in more than one table. For example, suppose we wish to know the CA TH classification of the domain containing residue 100 of PDB structure llga chain A. A possible SELECT statement is:

```
SELECT C, A, T, H
FROM PDB_DOM, PDB_DOM_SEG
WHERE PDB_DOM_SEG.PDBCODE = '1lga'
AND PDB_DOM_SEG.CHAIN = 'A'
AND PDB_DOM_SEG.RESNUM1 <= 100
AND PDB_DOM_SEG.RESNUM2 >= 100
AND PDB_DOM.PDBCODE = PDB_DOM_SEG.PDBCODE
AND PDB_DOM.CHAIN = PDB_DOM_SEG.CHAIN
AND PDB_DOM.DOMAIN = PDB_DOM_SEG.DOMAIN
```

producing result:

C	A	T	H
1	10	520	10

The query specifies both the PDB_DOM and PDB_DOM_SEG tables in the FROM clause: the query requires access to data in both tables. The first four conditions in the WHERE clause specify the conditions that a row in PDB_DOM_SEG must satisfy in order to figure in the result. For any row that does satisfy these conditions, the final three conditions in the WHERE clause specify the conditions which a row in PDB_DOM must itself satisfy in order for its C, A, T, H values to appear in the result, namely it must match the PDBCODE, CHAIN and DOMAIN values in the PDB_DOM_SEG row. SQL SELECT statements have many other features not illustrated by this example, but these features do not extend the fundamental way in which SELECT statements work. In essence, any SQL SELECT statement will retrieve data from one or more tables by using:

- a SELECT clause specifying the particular values which we wish to retrieve,
- a FROM clause to specify the tables which must be accessed, and
- an optional WHERE clause to specify conditions restricting the rows which we wish to retrieve from tables, and the conditions which specify how rows in different tables must match each other.

While SQL statements enable data to be retrieved from tables, developing a full application will usually require programming in a language such as Perl, Java or C++ for all the non-database tasks of the application. Hence, mechanisms have been developed to enable such languages to use SQL to access a database.

The main difficulty in using SQL in an ordinary programming language is that an SQL SELECT statement potentially results in many rows of data being retrieved to form the result, but programming languages are usually designed to handle a single row or record at a time. This mismatch is solved by use of an *SQL cursor* which enables row-by-row retrieval of results so a program can process a row at a time. The details of how to use SQL from a programming language depend on the particular language used. For Perl, a programming interface called *DBI* is widely used, as described by Descartes and Bunce. This is database independent, meaning it will work with most DBMSs including PostgreSQL, MySQL and Oracle.

A similar database-independent programming interface *IDBC* is used from within Java programs; White *et al.* provide a thorough tutorial and reference for JDBC. For C or C++ programs, an interface supporting *embedded SQL* is often used, in which SQL statements are interspersed with the ordinary C or C++ program statements. Embedded SQL is typically more dependent on the particular DBMS used than either Perl DBI or Java JDBC.

Designing a Database

There are two distinct but related aspects to designing a database, *logical design* and *physical design*. The former is concerned with designing the table structures which users of a relational database will see. The latter is concerned with designing the actual storage structures which underlie those table structures. This raises a fundamental aspect of databases which many database users remain unaware of: namely, the data model presented by a DBMS—a series of tables in the case of a relational database— may differ enormously from the storage structures maintained in reality by the DBMS to hold the data.

Logical design

Considering logical design first, there are usually many possible table designs for any given collection of data. This has introduced a single numeric identifer DOMAIN_ID to represent the combination of PDBCODE, CHAIN and DOMAIN. A new table PDB_DOM_ID has been introduced to record the relationship between the new DOMAIN_ID and the existing identifiers. The process of database logical design is that of choosing between

PDB_DOM_ID

DOMAIN_ID	PDBCODE	CHAIN	DOMAIN
1	1lga	A	1
2	1lga	A	2
3	1lga	B	1
4	1lga	B	2

PDB_DOM

DOMAIN_ID	C	A	T	H	S35	S95	S100
1	1	10	520	10	1	2	1
2	1	10	420	10	2	2	1
3	1	10	520	10	1	2	2
4	1	10	420	10	2	2	2

PDB_DOM-SEG

DOMAIN_ID	SEGMENT	RESNUM1	RESNUM2
1	1	11	144
1	2	269	299
2	1	145	268
2	2	300	339
3	1	11	144
3	2	269	296
4	1	145	268
4	2	297	339

Fig. 6.3. Table design 2.

different possible table designs such as seen here. One advantage is that domains are identified by a single numeric identifier. Hence, in queries, only this single identifier needs to be matched when combining rows from more than one table, rather than the values of three columns as was needed for the original design. Conversely, a disadvantage is that the query would involve three tables rather than the two or one of the alternative designs, and queries involving matching rows in multiple tables—*joins*—can be very slow.

Physical design

Turning to physical design, while a relational database appears to consist of tables, in reality the data are stored in a variety of storage structures designed to support quick retrieval of data. It is the aim of physical database design to choose the most appropriate storage structures for a database. Here we consider only the use of index storage structures. Indexes in databases work in a similar fashion to indexes in books. A reader wanting to find pages in a book related to a particular topic can simply work through the book page by page examining each until one is reached mentioning the topic of interest. For short books this may be acceptable, but for long books it is simply too time-consuming.

Similarly for a table in a database, if rows are sought containing a particular value or values, the DBMS can examine each row one-by-one, but this will be slow for large tables potentially holding millions of rows. Therefore, DBMSs support indexes on tables. An index identifies which rows contain particular values for one or more columns in a table.

1. An index on the combination of values of columns PDBCODE and CHAIN in PDB_DOM_SEG would enable a DBMS to retrieve rows very quickly in that table with particular values for the PDBCODE and CHAIN columns, as was needed for the query for the combination of values ('1lga', 'N).
2. An index on the combination of values of columns PDBCODE, CHAIN and DOMAIN in PDB_DOM would mean that once a row in PDB_DOM_SEG had been retrieved, the DBMS could very quickly retrieve matching rows in PDB_DOM with the same combination of PDBCODE, CHAIN and DOMAIN values.

The example illustrates the two main ways indexes are used by a relational DBMS: to quickly retrieve rows in a table with a given combination of values therein, and additionally to retrieve rows in one table matching rows in a second table as required in a join of the tables. Although indexes may give improved query performance, they do result in slower update performance since as values in table columns are updated, associated indexes must also be updated by the DBMS. This cost, as well as the space required to store the indexes, must be borne in mind when considering a physical design.

Overcoming Performance Problems

Once a database has been designed and loaded with data, it can be a shock to find that the SQL SELECT statements written to query the tables run very slowly indeed. While this might be expected for large tables containing millions of rows, even with tables containing thousands of rows it is very easy to write SQL queries which take many minutes and possibly hours before results are produced. Conversely, it can be a surprise that a SQL SELECT statement which does query tables containing millions of rows nonetheless produces results almost instantaneously.

1. First, examine the SQL query itself. Has it been written correctly to return the results you intended? In particular, check that the query does not refer to more tables than are necessary in the FROM clause, or that the WHERE clause is missing a join condition specifying how the values in two tables should match.
2. Second, make sure the physical design of the tables is appropriate. It was noted that indexes enable a relational DBMS to retrieve rows quickly. Have the indexes been defined to support the problem query? Are there other physical storage structures supported by the DBMS which would help the query run faster?
3. Third, if the physical design seems to be appropriate, is the DBMS choosing a sensible strategy to retrieve rows? When a DBMS is given an SQL query to execute, it considers alternative ways to produce the required results. If, for example, it needs to do a join retrieving rows in one table matching rows in a second table, it has a number of options. It could simply retrieve the rows in the first table one by one, and check each against every row in the second table to find matches. Alternatively, it could use an index to find matching rows in the second table. Yet another alternative would be for it to sort both tables on the values being matched, and then scan each sorted table in parallel looking for matches. The DBMS is carrying out *query optimization* in choosing between the many options it has for producing the results of a query, and sometimes it does not make the best choice. Many DBMS products have tools to enable a user to view the strategy used by the DBMS to execute a query - the *execution plan*—and provide ways enabling a user to change the execution plan so that the DBMS executes the query more quickly. This might involve writing the SQL query in a slightly different way, or running a utility which updates the information the query optimizer uses to choose an execution plan, or changing one of the parameters controlling the

detailed way the DBMS works. The details will be specific to each DBMS.

If none of these results in the query running quickly enough, more drastic solutions must be considered:

1. Should the computer be upgraded with more memory or processors? DBMS products are notoriously greedy in their use of computer resources.
2. Should the logical design of the database be changed? For example, should the design be changed? The problem query would then have to retrieve data from only one table rather than two, so removing the table join which could be causing the performance problems. The implications of changing a table design are considered further below.
3. If all else fails, then there is no alternative but to accept that it is not feasible to run the query against the database. In that situation, one would look for a simpler, more efficient query to extract data from the database, and then manipulate that data outside the database using a conventional language such as Perl or C++.

The option of changing the logical design of a database was noted above as one way of removing performance problems with some queries. Of course, it is highly desirable that the most appropriate logical design is chosen in the light of expected queries before the database is created. A change to the logical design of an existing database is not a task to be undertaken lightly, since that may then force changes to be made to all the SQL queries in applications retrieving data from the original tables. DBMS products reduce this problem by allowing the definition of *view* tables, which look like the original tables but do not exist as tables stored in their own right. Instead, view tables are defined by SQL queries which are run by the DBMS when needed, to give the impression that the view tables are actually stored.

In this way existing application programs may be able to continue with SQL queries as originally written, but those queries now reference view tables giving the appearance of the original table design. Since what appears to be a single view table may in reality be produced by an SQL query joining a number of real tables, retrieving data from a view table may be slow given the joins involved. Some DBMS products support *materialized views* which, like ordinary view tables, are tables defined in terms of an SQL query, but in which the data for the resulting table are explicitly stored as a real table. As changes are made to the tables on which the materialized view is defined, the DBMS can then sometimes automatically make corresponding changes to the stored materialized view table.

Accessing Data from Remote Sites

It was noted that the data about any biological object are almost certainly held at many sites rather than in a single database. Hence, scientists typically need to be able to access data from many sites in order to carry out their work. Traditionally, this has meant downloading files from remote sites and writing programs to access those local copies of files, which has added to the problems of file-based data. With the advent of the World Wide Web, it has become common for web-based interfaces to be made available for browsing and searching data at remote sites, but in order for a scientist then to write programs to process such data, there has often still been no alternative other than to download the data to a local machine. DBMS-based databases have a number of features to make it easier for programs to access data held remotely.

1. First, DBMS software usually follows a *client-server* model. This means that the database and the DBMS software for managing it can be at one *server* site, while an application program and the DBMS software needed to enable that program to access the database can be at a separate *client* site. The database file is at a server site, each program may be at a different client site, and the DBMS software itself runs partly at the server site and partly at each client site.
2. A further level of flexibility is provided by a *distributed database*. This enables the database data to be spread across a number of sites, so what appears to a program to be a single database may in fact be a series of databases distributed across multiple server sites.

The techniques just described may be sufficient if all the remote data are held in the databases managed by a particular DBMS product. Even if the remote data are held in databases managed by different DBMS products, there is generally no problem in writing a program to access those data. Both the Perl DBI and Java JDBC are designed to work with multiple databases managed by different DBMS products. The Java JDBC interface can also be used from a client program without any DBMS software being installed on the client machine: the DBMS software needed is downloaded automatically from the server site. Another useful feature of Java JDBC is that it enables data held in files outside any database to be accessed nonetheless as though they were stored in relational database tables. The real problem with accessing remote data stored at multiple sites managed by different DBMS products or outside any database is that the data are no longer integrated and shared.

Each database is completely separate and so, once more, with each box being a database rather than an individual file. All the problems of that model then reappear. One approach to dealing with such problems is to build specialized software to cross-link data in different biological data sources, together with facilities to enable data from those sources to be retrieved in response to a query. *SRS* (Sequence Retrieval System) was developed at EMBL to achieve this for molecular biology data, and provides facilities to link sequence data (nucleic acid, EST and protein) together with wider data sources such as protein structure and bibliographic data resources.

SRS supports querying of those resources from both Web-based interfaces and a command line interface suitable for use from programs. Some general technologies have been developed to enable data from multiple remote sites nonetheless to be treated as a single, integrated resource. In particular:

1. *Data warehouses* are formed by extracting data from primary data sources and then holding the data in a database specifically designed to allow the relationships between the different data from the primary sources to be modeled and searched.
2. *Mediator* software is designed to run between the server software of remote data server sites and the application programs of client sites. A client program is given the illusion that the data are all stored according to a single common data model. This is achieved by the mediator software intercepting client requests for data, translating those requests into the format required by a particular data server, and then transforming the data retrieved from a server site into a form consistent with the common data model for transmission

back to the client program. The term *federated database* is sometimes used to refer to a system consisting of a number of server sites managing independent databases controlled by different DBMS products, but for which a single common data model is also supported to enable the databases to be accessed as a single resource.

3. *XML technology* enables data from a remote data server to be retrieved with tags which explicitly represent the meaning of the data in a form suitable for processing by computer programs. XML has become particularly important for enabling data to be passed between computer programs over the World Wide Web.

In recent years, DBMS products have become increasingly sophisticated in their support for these kinds of technologies. A DBMS no longer just manages data in tables in a single local database; increasingly it provides access to data stored locally and remotely, the data being stored in conventional databases, or in files external to the database, or on the World Wide Web.

Challenges Arising from Biological Data

The first problem was the many differing formats in which biological data are found, ranging from the regular structured data of GenBank or PDB files, to data of complex structure such as pathway data, or unstructured data such as literature abstracts. This is a problem for a DBMS which only supports a simple relational table model of data. However, many DBMSs have wider capabilities such as the following.

- Object-oriented DBMSs do have facilities for defining objects having a complex structure, together with the operators for storing and retrieving those objects. Hence, sequence or pathway objects could, for example, be defined and manipulated with such a system. Other DBMS products, known as *object-relational* DBMSs, have the ability to define such objects while retaining tables as the primary representation of data in the database.
- Some DBMS products have special facilities for holding and managing unstructured text data such as are found in biomedical literature abstracts.
- *Semi-structured database* techniques have been developed to handle data which, while not wholly unstructured, do not have the regular structure making them suitable to hold in

table form. Currently, XML techniques are often used to handle such data.

The second problem was that there is often no single biological object to which data refer: individual users of the data will have their own perspectives. This is equally the case in considering data concerned with biological functions and mechanisms.

- This remains a problem with many DBMS products. The relational model provides a fixed table representation of the data although view tables can go some way to providing alternative views of the data. More flexible views can be obtained by holding the data in a semi-structured form, with table views being generated as required.

The third problem was the many incompatible formats and nomenclatures for biological data. The key to overcoming this is agreement on standards for representing biological data and their semantics. There are a number of initiatives in this area. For example:

- The *GO* (*Gene Ontology*) consortium has developed a controlled vocabulary for terms describing gene products from three perspectives: molecular function, biological process and cellular component. The ontology explicitly defines the relationship between more specific and more general terms in graph structures for each of these three perspectives. The ontology may be obtained in XML format, and also exists as a relational database which may be queried from a Web-based interface.
- A very large number of XML-based definitions and other resources have been published. These range from XML representations of specific data resources such as GO, InterPro and PIR, to proposals for XML representations of sequence data (BIOML, BSML), BLAST output (BlastXML) and expression data (MAGE-ML), as well as more general XML definitions for ontology specification (DAML+OIL). One example of XML being used in a practical way is DAS (Distributed Annotation System) which uses an XML representation of sequence annotation information to enable integration of annotations retrieved from multiple sites.
- In the object database world, the OMG LSR (Object Management Group Life Sciences Research) provides a focus

for similarly defining standards and interfaces supporting a common format for biological data.

The final problem was the fact that data about any given biological object are likely to be held at many different sites, each with a different approach to coping with the previous problems.

- Techniques for retrieving data from remote sites are well established. Once again, the problems are those of being able to treat those data as a single integrated resource, and the key is agreement on standards for representing data and their semantics.

Conclusions

The data management techniques outlined in this chapter will undoubtedly be developed further to enable distributed biological data to be used as an integrated resource. Many of the most difficult outstanding problems arise from the semantic integration of such data rather than any technical problems in its storage and access as such.

Currently the World Wide Web enables distributed data to be accessed and displayed in a browser, but a program cannot extract the meaning of the data. The term *semantic web* is used to describe what data management techniques will increasingly support: ways of enabling distributed data to be accessed in a form whereby the semantics of the data are explicit and may be extracted and exploited by computer programs. In addition, the computer programs themselves will be distributed resources on the Web, rather than being written by individual scientists to run on local machines. This is the aim of *grid* computing: turning the Web into integrated data and computational resources which a scientist can plug into in the same way that the electricity grid can be plugged into when electrical power is required.

7

Calcium Waves and Sparks

In this chapter we shall discuss a variety of intracellular Ca^{2+} wave phenomena, but always from the perspective that the distance scales of interest are large enough that Ca^{2+} transport is well-modeled by conservation equations based on a continuum description of matter. Although recent experimental and theoretical work suggests that the macroscopic behavior of propagating Ca^{2+} waves (*e.g.*, wave speed) may depend in subtle ways on the density and distribution of intracellular Ca^{2+} release channels, we postpone consideration of intracellular heterogeneities such as clusters of Ca^{2+} release channels until later in the chapter. This makes sense because both the mathematics and simulation methods used to study nonlinear wave propagation in *homogeneous* media are simpler than the *heterogeneous* case. This simplicity should facilitate the development of intuition regarding nonlinear wave propagation. Throughout the chapter a recurring theme will be the manner in which Ca^{2+} buffers, through their important association with free Ca^{2+}, can influence wave phenomena dependent on diffusion. The chapter concludes with calculations of localized Ca^{2+} elevations due to intracellular Ca^{2+} release, *i.e.*, Ca^{2+} "puffs" or "sparks," elementary events that sum to produce Ca^{2+} waves.

Microfluorometric Measurements

Experimental observations of such intracellular Ca^{2+} dynamics are often made using microfluorimetry, an experimental technique that involves loading Ca^{2+} indicator dyes into cells and instrumentation that optically excites these indicators and measures emission. These Ca^{2+} indicator dyes are themselves Ca^{2+} buffers (often highly mobile) that can potentially affect intracellular Ca^{2+} signaling. For example, the differential fraction of free to bound cytosolic Ca^{2+}, denoted by $f_i([Ca^{2+}]_i)$, will be determined by both exogenous as well as endogenous Ca^{2+} buffers.

Although a measured fluorescence signal is only indirectly related to the dynamics of intracellular Ca^{2+}, it is relatively straightforward to determine the free Ca^{2+} concentration during a cell-wide Ca^{2+} response using the time course of measured fluorescence. If the Ca^{2+} and indicator dye concentrations are homogeneous throughout the cell, the equilibrium relation

$$[\text{CaB}] = \frac{[\text{Ca}^{2+}]_i\,[\text{B}]_\text{T}}{\text{K}+[\text{Ca}^{2+}]_i} \quad ...(1)$$

is valid as long as $[Ca^{2+}]_i$ changes slowly compared to the equilibration time of the buffers. Because this equilibration time is on the order of milliseconds, this condition is usually satisfied for global Ca^{2+} responses, which occur with a time scale of seconds or tens of seconds.

If the equilibrium relation (1) accurately describes the relationship between the concentration of bound indicator dye ([CaB]) and free Ca^{2+} ($[Ca^{2+}]_i$), then it is a simple matter to "backcalculate" the free Ca^{2+} concentration as a function of time. For a single excitation wavelength measurement (*e.g.*, using a nonratiometric dye such as fluo-3 at low concentration), we can idealize the indicator fluorescence as the sum of two components,

$$F = \eta_B[\text{B}] + \eta_{CaB}[\text{CaB}], \quad ...(2)$$

where η_B and η_{CaB} are proportionality constants for free and bound dye, respectively. When $\eta_B < \eta_{caB}$. the maximum and minimum observable fluorescences are given by $F_{min} = \lim_{[Ca^{2+}]i\to 0} F = \eta_B[B]_T$ and $F_{max} = \lim_{[Ca^{2+}]i\to\infty} = \eta_{CaB}[B]_T$. Using the equilibrium relation it can be shown that

$$[\text{Ca}^{2+}]_i = K\frac{[\text{CaB}]}{[\text{B}]} = K\frac{F - F_{min}}{F_{max} - F}. \quad ...(3)$$

If only the Ca^{2+}-bound indicator fluoresces strongly, then $F \approx \eta_{caB}[\text{CaB}]$, and a slightly simpler expression results from substituting $F_{min} = 0$ in (3).

It should be noted that the validity of (3) relies on the stability of instrument sensitivity, optical path length, and dye concentration between measurements of F, F_{min}, and F_{max}. Because determining F_{min} and F_{max} usually involves titrating the indicator released from lysed cells, this is difficult to achieve in practice.

In whole-cell Ca^{2+} measurements, fluorescence intensities can be measured at two excitation wavelengths (λ and λ') using indicator dyes such as fura-2 Such *mtiometric* measurements can be related to the underlying free Ca^{2+} signal by supplementing (2) with

$$F' = \eta'_B[B] + \eta'_{CaB}[CaB], \qquad ...(4)$$

where the primes indicate the second excitation wavelength. Using the first equality of (3), the fluorescence ratio $R = F/F'$ can be inverted to give

$$[Ca^{2+}]_i = K\frac{\eta_B - \eta'_\beta R}{\eta'_{CaB}R - \eta_{CaB}} = K\left(\frac{R - R_{min}}{R_{max} - R}\right)\left(\frac{\eta'_\beta}{\eta'_{CaB}}\right), \qquad ...(5)$$

where for the second equality we use $R_{min} = \lim_{[Ca^{2+}]_i \to 0} F/F' = \eta_B/\eta'_B$ and $R_{max} = \lim_{[Ca^{2+}]_i \to \infty} F/F' = \eta_{CaB}$. If λ' is chosen to be a wavelength at which the calibration spectra at different Ca^{2+} concentrations cross one another, then $\eta'_B \approx \eta'_{CaB}$, and the last factor in (5) is eliminated. An advantage of the ratiometric method is its insensitivity to changes in dye concentration and instrument sensitivity between measurements.

Model of the Fertilization Calcium Wave

When mature *Xenopus laevis* oocytes (eggs) are loaded with an indicator dye (*e.g.*, Ca^{2+}-green dextran) and stimulated by the fusion of sperm, a propagating wave of intracellular Ca^{2+} release can be observed by backcalculating the free Ca^{2+} concentration ($[Ca^{2+}]_i(x, t)$ from a time-dependent fluorescence signal ($F(x, t)$) according to (3) or (5). This fertilization Ca^{2+} *wave* is an important step in early development. It triggers the fusion of cortical granules (vesicles) with the plasma membrane, a process that initiates the raising of the viteline envelope and a long-lasting block to polyspermy. The cell divisions that initiate development of *Xenopus* begin only after the fertilization Ca^{2+} wave has propagated throughout the entire cell. The eggs of many species, from starfish to mammals, exhibit propagating Ca^{2+} waves upon fertilization.

Because fertilization Ca^{2+} waves still occur when the extra-cellular medium is nominally Ca^{2+} free, the phenomenon appears to be largely independent of Ca^{2+} influx. On the other hand, the fertilization Ca^{2+} wave absolutely requires functional IPs receptors. When IPs-mediated Ca^{2+} release is blocked by any one of several experimental manipulations, the fertilization Ca^{2+} wave is not observed.

We expect Ca^{2+} diffusion to play an important role in the fertilization Ca^{2+} wave because *Xenopus luevis* eggs are large (approximately 1.2 mm in diameter). Indeed, the Ca^{2+} concentration within the egg depends very much on both spatial position and time. For this reason a whole cell model of this phenomenon would be deficient, and instead researchers mathematically describe the fertilization Ca^{2+} wave in the *Xenopus* egg and other cell types using a combination of spatial modeling and whole-cell modeling approaches.

Here we introduce such a spatial whole-cell model of the fertilization Ca^{2+} wave by recalling that the Li-Rinzel reduction of the DeYoung-Keizer model of the IPsR can be combined with sigmoidal SERCA pump kinetics and a passive ER leak to create a whole-cell model of Ca^{2+} handling in pituitary gonadotrophs. Using a variation of this model to represent Ca^{2+}-induced Ca^{2+} release (CICR) and reuptake by the endoplasmic reticulum (ER), and account for Ca^{2+} diffusion in both the cytosol and ER by writing the following reaction-diffusion system:

$$\frac{\partial[\mathrm{Ca}^{2+}]_i}{\partial t} = f_i[D_i\nabla^2[\mathrm{Ca}^{2+}]_i + j_{\mathrm{IP_3R}} + j_{\mathrm{LEAK}} - j_{\mathrm{SERCA}}], \qquad ...(6)$$

$$\frac{\partial w}{\partial t} = [w_\infty([\mathrm{Ca}^{2+}]_i, [\mathrm{IP}_3]) - w]/\tau, \qquad ...(7)$$

$$\frac{\partial[\mathrm{Ca}^{2+}]_{\mathrm{ER}}}{\partial t} = f_{\mathrm{ER}}[D_{\mathrm{ER}}\nabla^2[\mathrm{Ca}^{2+}]_{\mathrm{ER}} - (\bar{V}_i/\bar{V}_{\mathrm{ER}})(j_{\mathrm{IP_3R}} - j_{\mathrm{LEAK}} + j_{\mathrm{SERCA}})], \qquad ...(5.8)$$

where $j_{\mathrm{LEAK}} = v_{\mathrm{LEAK}}\,[\mathrm{Ca}^{2+}]_{\mathrm{ER}} - ([\mathrm{Ca}^{2+}]_i)$, $j_{\mathrm{IP_3R}} = v_{\mathrm{IP_3R}} m_\infty^3 w^3\ ([\mathrm{Ca}^{2+}]_{\mathrm{ER}} - [\mathrm{Ca}^{2+}]_i)$, $j_{\mathrm{SERCA}} = v_{\mathrm{SERCA}}[\mathrm{Ca}^{2+}]_i^2/([\mathrm{Ca}^{2+}]_i^2 + K_{\mathrm{SERC}}^2)$, and $\bar{V}$'s are the volumes as before. In these equations, w is the fraction of IP_3Rs not inactivated and the open probability of the IP_3Rs is given by $P_O = m_\infty^3 w^3$ where the fraction of activated IP_3Rs (m_∞^3) is assumed to be an instantaneous function of Ca^{2+} and IPs concentrations,

$$m_\infty^3 = \left(\frac{[\mathrm{IP}_3]}{[\mathrm{IP}_3 + d_1}\right)^3 \left(\frac{[\mathrm{Ca}^{2+}]_i}{[\mathrm{Ca}^{2+}]_i + d_5}\right)^3. \qquad ...(9)$$

Although the Li-Rinzel reduction of the DeYoung-Kiezer model gives a time constant of IP_3R inactivation that is dependent on Ca^{2+} and IP_3 concentration, $\tau([Ca^{2+}]_i, [IP_3])$, for simplicity we will assume that τ is constant (2 sec).

Note that in (6)-(8) the maximum conductance of IP_3-mediated Ca^{2+} release (v_{IP_3R}), passive leak rate (v_{LEAK}), and maximum rate of ATP-dependent reuptake (v_{SERCA}) are constants. Thus, the model assumes homogeneous Ca^{2+} release and reuptake dynamics, that is, a uniform and high density of inositol 1,4,5-trisphosphate (IPs) receptors and sarco-endoplasmic reticulum Ca^{2+}-ATPases. Also note that (7) includes no diffusion term for the simple reason that IPsRs (and thus the gating variable representing their inactivation) has a fixed spatial location and does not diffuse. Throughout this chapter we will assume that $[IP_3]$ is uniform and constant.

Including Calcium Buffers in Spatial Models

In this spatial whole-cell model given by (6)-(8), the quantities fi and fea multiply both reaction and diffusion terms. These factors scale the Ca^{2+} release and reuptake rates to account for the proportion of Ca^{2+} bound to buffer in both cytosolic and ER compartments.

Note that the coefficients of the Laplacians in (6) and (8) imply constant *effective diffusion coefficients* for the cytosol and ER given by $D_i^{\text{eff}} = f_i D_i$ and $D_{\text{ER}}^{\text{eff}} = f_{\text{ER}} D_{\text{ER}}$, both of which are expected to be much smaller than the free Ca^{2+} diffusion coefficient. This assumption is here made for convenience and is strictly true only for low-affinity Ca^{2+} buffers.

To see that this is the case, consider how one Ca^{2+} buffer (perhaps representing a Ca^{2+} indicator dye) would be added to the fertilization Ca^{2+} wave model. The association and dissociation of Ca^{2+} with the indicator dye would contribute bimolecular reaction terms in the cytosolic compartment, giving

$$\frac{\partial[\text{Ca}^{2+}]_i}{\partial t} = D_C \nabla^2 [\text{Ca}^{2+}]_i + R + j_T, \qquad \text{...(10)}$$

$$\frac{\partial[\text{B}]}{\partial t} = D_B \nabla^2 [B] + R, \qquad \text{...(11)}$$

$$\frac{\partial[\text{CaB}]}{\partial t} = D_{CB} \nabla^2 [\text{CaB}] - R, \qquad \text{...(12)}$$

where D_C, D_B, and D_{CB} are diffusion constants for free Ca^{2+}, free buffer, and bound buffer, respectively,

$$R = k^{+}[B][Ca^{2+}]_i + k^{-}[CaB],$$

and $j_T = j_{IP_3R} + j_{LEAK} - j_{SERCA}$ is the sum of all Ca^{2+} fluxes into and out of the ER.

Notice that if the diffusion of buffer doesn't depend on whether or not Ca^{2+} is bound ($D_B \approx D_{CB}$), then (11) and (12) can be summed to give 9[B]r, where $[B]_T = [B] + [CaB]$ is the total buffer concentration profile (free plus bound). This equation implies that if $[B]_T$ is initially uniform (not a function of position), it will remain uniform for all time. We can thus eliminate (11) and obtain a reduced system given by (10) and (12) with

$$R = -k^{+}([B]_T - [CaB])[Ca^{2+}]_i + k^{-}[CaB].$$

Effective Diffusion Coefficient

If the buffer reactions are rapid with respect to the diffusion, it is possible to simplify our model further using the rapid buffer approximation. Assuming local equilibrium (*i.e.*, chemical equilibrium at every point in space), we can write an equilibrium expression for total cell Ca^{2+} concentration:

$$[Ca^{2+}]_T = [Ca^{2+}]_i + [CaB] = [Ca^{2+}]_i + \frac{[Ca^{2+}]_i [B]_T}{[Ca^{2+}]_i + K}.$$

Next, we eliminate reactions terms (and the fast time scale) by adding (10) and (12), to give

$$\frac{\partial [Ca^{2+}]_T}{\partial t} = \frac{\partial [Ca^{2+}]_i}{\partial t} + \frac{\partial [CaB]}{\partial t} = D_C \nabla^2 [Ca^{2+}]_i + D_B \nabla^2 [CaB] + j_T \quad \text{....(13)}$$

The reader can confirm that substituting $[Ca^{2+}]_T$ and $[CaB]$ in terms of $[Ca^{2+}]_i$ and taking derivatives gives the so-called rapid buffer approximation,

$$\frac{\partial [Ca^{2+}]_i}{\partial t} = \beta \left((D_C + D_{B\gamma}) \nabla^2 [Ca^{2+}]_i - \frac{2\gamma D_B}{K + [Ca^{2+}]_i} (\nabla^2 [Ca^{2+}]_i)^2 + j_T \right),$$

where $\gamma = K[B]_T/(K + [Ca^{2+}]_i)^2$ and $\beta = 1/(1 + \gamma)$. Although this equation may appear ominous, note that using the rapid buffer approximation we have been able to eliminate the two extra equations (for [B] and [CaB]) needed to explicitly account for an indicator dye in the cytosolic compartment.

Another valuable feature of (14) is that it provides some insight into the effect of Ca^{2+} buffers on Ca^{2+} transport in cells by allowing us to identify the Ca^{2+}-dependent effective diffusion coefficient,

$$D^{\text{eff}} = \beta(D_C + \gamma D_B)$$

Furthermore, if this buffer has low affinity so that $[Ca^{2+}]_i \ll K$ and $y \approx [B]_T/K$, then as promised at the beginning of this section this diffusion coefficient is approximately constant:

$$D^{\text{eff}} \approx \frac{K}{K+[\mathrm{B}]_{\mathrm{T}}}\left(D_C + \frac{[\mathrm{B}]_{\mathrm{T}}}{K}D_B\right).$$

For simplicity the remainder of this chapter assumes a rapid and low-affinity buffer.

Simulation of a Fertilization Calcium Wave

Now that we have justified the reduced effective diffusion coefficients used in the fertilization Ca^{2+} wave model, let us further simplify (6)-(8) by assuming that ER depletion is minimal, *i.e.*, $[Ca^{2+}]_{ER} = c_{ER}$ is approximately constant. We will also assume that the dynamics of $[Ca^{2+}]_i$ are much slower than the gating variable for Ca^{2+} inactivation of the IP_3R, so that w is well approximated by $w_\infty([Ca^{2+}]_i)$. With these assumptions, (6) and (7) reduce to the single-variable reaction-diffusion equation

$$\frac{\partial c}{\partial t} = D\nabla^2 c + f(c), \qquad ...(15)$$

where we have written $c = [Ca^{2+}]_i$, $D = f_i D_i$, and

$$f(c) = f_i\left([v_{\mathrm{LEAK}} + v_{\mathrm{IP_3R}}(c)][c_{\mathrm{ER}} - c] - \frac{v_{\mathrm{SERCA}}c^2}{c^2 + K_{\mathrm{SERCA}}^2}\right), \qquad ...(5.16)$$

and where

$$v_{\mathrm{IP_3R}}(c) = [m_\infty(c)w_\infty(c)]^3.$$

Using parameters, Figure 7.1 shows that $f(c)$ is a cubic function with three zeros. Equilibrium point 1 represents a stable resting state at basal $[Ca^{2+}]_i$ (ER replete), while equilibrium 3 is a stable resting point at high $[Ca^{2+}]_i$ (ER empty). Equilibrium 2 lies between points 1 and 3, is unstable, and corresponds to a threshold for Ca^{2+}-induced Ca^{2+} release.

Simulation of a Traveling Front

We follow previous section to simulate our minimal model of a fertilization Ca^{2+} wave given by (15). Assuming a one-dimensional geometry ($0 \le x \le L$), we discretize space so that $C_j(t)$ is an approximation to $c(x_j, t)$, where $x_j = j\Delta x$, $0 \le j \le j$, and $\Delta x = L/J$. In this way we arrive at the system of ODEs

$$\frac{dc_0}{dt} = D\frac{c_1 - c_0}{(\Delta x)^2} + f(c_0),$$

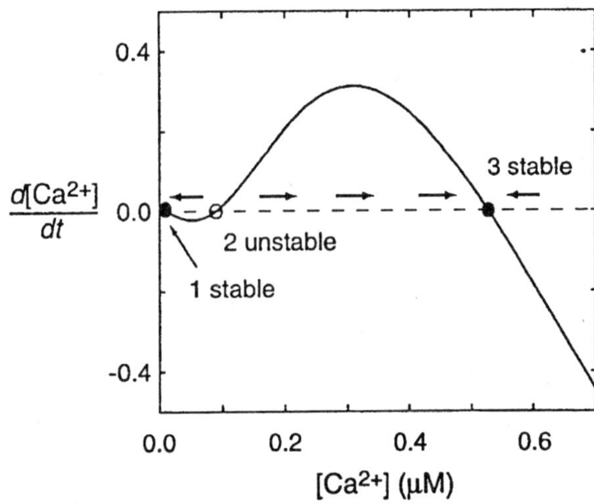

Fig. 7.1. Using parameters, the cubic rate function dc/dt = f(c) of the fertilization Ca^{2+} wave model, leads to a bistable phase portrait (x-axis).

$$\frac{dc_j}{dt} = D\frac{c_{j+1} - 2c_j + c_{j-1}}{(\Delta x)^2} + f(c_j), \qquad \text{...(5.17)}$$

$$\frac{dc_J}{dt} = D\frac{c_{J-1} - c_J}{(\Delta x)^2} + f(c_J),$$

where the equations for c_0 and c_J are found using a discretized version of no-flux boundary conditions $\partial c/\partial x|_{x=0,L} = 0$ to specify values at the "ghost points" $c_{-1} = c_0$ and $c_{J+1} = c_J$.

Figure 7.2 shows a simulation of a fertilization Ca^{2+} wave calculated using (18). Our assumed one dimensional geometry suggests that we interpret the calculation as a propagating planar signal ($[Ca^{2+}]_i$ not a function of y or z, but only x). Initial conditions are chosen so that elevated $[Ca^{2+}]_i$ in a small region triggers a wave traveling leftwards at a velocity of 3.45 μm/s. The basal $[Ca^{2+}]_i$ before the front passes is given by equilibrium point 1 (0.01 μM), and after the front passes elevated $[Ca^{2+}]_i$ is given by equilibrium point 3 (0.53 μM). Because $[Ca^{2+}]_i$ doesn't return to basal values after the wave passes, this type of traveling wave is referred to as a *propagating front*.

In the fertilization Ca^{2+} wave model, we could say that the physiological state of the cell cytoplasm is bistable, that is, if the spatial component of the model is neglected (for example, by setting $D = 0$ in (15)), the scalar ODE $(dc/dt - f(c))$ that remains

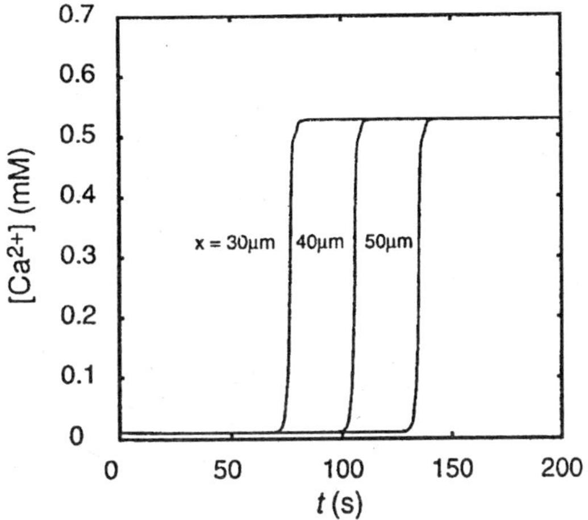

Fig. 7.2. Simulation of a fertilization Ca^{2+} wave following (eq. 15) results in a traveling front.

exhibits bistability. Alternatively, imagine walling off (or lifting out) a small region of the spatial model and investigating its dynamics. If the region is small enough, the time scale for diffusion in the isolated region ($T = L^2/D$) will be much faster than the dynamics associated with the reaction terms $f(c)$. Thus, the Ca^{2+} profile will be approximately uniform ($\partial c/\partial x = 0$), and the region will behave as a *compartmental model* in which diffusion no longer plays a role. The propagating front observed in the spatial model, Figure 7.2, is possible precisely because any such isolated compartmental model would exhibit bistability. Indeed, if the physiological state of the cytosol were not bistable, the basal and elevated $[Ca^{2+}]_i$ at the wave front and back could not persist. We will see below that spatial phenomena observed in reaction-diffusion models can be categorized by the qualitative dynamics of the ODEs obtained by assuming that all profiles are uniform.

We can analyze the fertilization Ca^{2+} wave model (15) by looking for a (rightward) traveling wave solution $c(x, t) = C(x - vt) = C(z)$. Making this substitution as well as $\partial C/\partial x = (dC/dz)(\partial z/\partial x) = dC/dz$ and $\partial C/\partial t = (dC/dz)(\partial z/\partial t) = -v dC/dz$ into this reaction-diffusion equation gives

$$-v\frac{dC}{dz} = D\frac{d^2C}{dz^2} + f(C), \qquad \text{...(18)}$$

which can be written as the first-order system

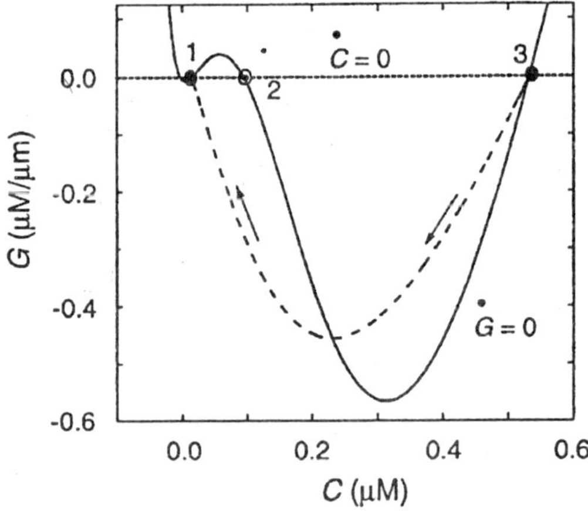

Fig. 7.3. The reaction-diffusion equation model of the fertilization Ca^{2+} wave, (eq. 18), can be transformed into traveling wave coordinates ($z = x - vt$) resulting in a first-order system of ODEs (eq. 19 and 20).

$$\frac{dC}{dz} = G, \quad \text{...(5.19)}$$

$$\frac{dG}{dz} = -\frac{1}{D}[vG + f(C)]. \quad \text{...(5.20)}$$

The phase portrait in Figure 7.3 is a plot of the rate function $C' = G$ versus Ca^{2+} concentration C. The reader can numerically demonstrate the existence of a well-behaved heteroclinic orbit when v = 4.96 µm/s. This heteroclinic orbit does not exist for smaller or large values of v and corresponds to the traveling front presenting in Figure 7.2.

Intitively, we expect the speed of traveling front solutions such as those presented in Figure 7.2 and Figure 7.3 to depend on the diffusion coefficient D. Indeed, when the rate function $f(c)$ is a cubic polynomial, the wave speed is proportional to the square root of the diffusion coefficient. Although in the fertilization Ca^{2+} wave model $f(c)$ is not a polynomial, we can demonstrate the same principle by noting that the substitution $D^* = \alpha^2 D$ in (18) can be offset by scaling the wave speed ($v^* = \alpha v$). To see this, we define a scaled wave coordinate $z^* = x - v^* t$ that implies $-v^* \, dC/dz^* = -v dC/dz$ or $dC/dz^* = a^{-1} dC/dz$. Thus, any solution of (18) is also a solution of

$$v^* \frac{dC}{dz^*} = D^* \frac{d^2C}{dz^{*2}} + f(C).$$

The reader can repeat the simulations and/or numerically confirm that the traveling front velocity is proportional to the square root of the diffusion coefficient. Interestingly, exogenous buffers have been shown to alter the speed of Ca^{2+} waves in mature *Xenopus* oocytes, an effect that may be due to changes in the effective diffusion coefficient for Ca^{2+}.

The astute reader may have noticed that we have reported two different wave velocities for two different calculations of the same traveling front. Recall that Figure 7.2 used the method of lines to simulate the fertilization Ca^{2+} wave model (15) and that we observed a speed of 3.45 μm/s. On the other hand, the first-order system of ODEs resulting from a transformation into traveling wave coordinates, (19)-(20), has a heteroclinic orbit that connects the two stable equilibria for wave velocity of v = 4.96 μm/s. The suppression of the wave speed in the calculation using the method of lines is due to *discretization error.*

Calcium Waves in the Immature Xenopus Oocycte

Although we have begun our discussion of models of intracellular Ca^{2+} waves by focusing on traveling fronts in mature *Xenopus laevis* eggs, IP_3-dependent Ca^{2+} responses occur in many cell types and take various forms. It is instructive to note here that Ca^{2+} wave phenomena in the *immature X. laevis* oocyte (such as the spiral Ca^{2+} wave) are qualitatively distinct from the traveling fronts discussed above. Although there are many differences between immature and mature *Xenopus* oocytes/eggs, from a dynamical point of view these different Ca^{2+} wave phenomena are the result of different underlying dynamics of Ca^{2+} handling that could be distinguished even in the absence of diffusion. While our model of the fertilization Ca^{2+} wave demonstrated how bistable cellular dynamics can lead to a traveling front, the ER of the immature *Xenopus* oocyte is excitable (and may even be oscillatory) as opposed to bistable.

Figure 7.4 presents phase planes for the reaction terms of the fertilization Ca^{2+} wave model (6)-(7), which illustrate the qualitative difference between excitable and bistable dynamics. To make this comparison, we relax our assumption that the dynamics of $[Ca^{2+}]_i$ are much slower than the gating variable for Ca^{2+} inactivation of the IPsR (so that w is once again given by (7) rather than $w_\infty([Ca^{2+}]_i)$). The model now has two dynamic variables ($[Ca^{2+}]_i$ and w), the nullclines of which are shown in

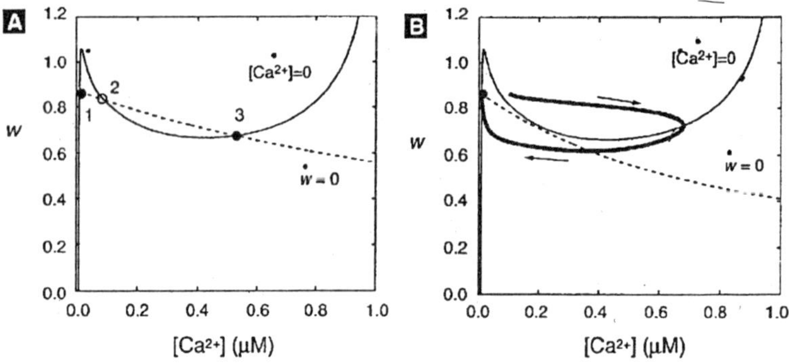

Fig. 7.4. The (w, $[Ca^{2+}]_i$) phase plane for the reaction terms of the fertilization Ca^{2+} wave model can exhibit bistability (A) or excitability (B) with a change of IP_3R parameters.

panel A of Figure 7.4. The intersections of these nullclines imply three equilibria (as before, points 1 and 3 are stable, while 2 is unstable). Conversely, panel B presents the (w, $[Ca^{2+}]_i$) phase portrait with adjusted IP_3R parameter values (Ca^{2+}-dependence of inactivation). The reaction terms of the model now represent excitable cytoplasm. A small perturbation in Ca^{2+} concentration results in a long excursion of the trajectory representing Ca^{2+} release from the ER, IP_3R inactivation, and ATP-dependent Ca^{2+} reuptake.

Simulation of a Traveling Pulse

A simulation of Ca^{2+} wave phenomena mediated by ER Ca^{2+} excitability can be implemented in a manner similar to (18). To do this we discretize the gating variable $w_j \approx w(x_j, t)$ at $J + 1$ mesh points ($0 \leq j \leq J$) representing the fraction on noninactivated IP_3Rs at each location in space ($x_j = j\Delta x$). Of course, there will be no Laplacian term in the equations for w_j because the IP_3Rs (and thus their state) do not diffuse. The model now takes the form

$$\frac{dc_0}{dt} = D\frac{c_1 - c_0}{(\Delta x)^2} + g(c_0, w_0), \qquad \frac{dw_0}{dt} = [w_\infty(c_0) - w_0]/\tau,$$
$$\frac{dc_j}{dt} = D\frac{c_{j+1} - 2c_j + c_{j-1}}{(\Delta x)^2} + g(c_j, w_j), \quad \frac{dw_j}{dt} = [w_\infty(c_j) - w_j]/\tau, \qquad ...(21)$$
$$\frac{dc_J}{dt} = D\frac{c_{J-1} - c_J}{(\Delta x)^2} + g(c_J, w_J), \qquad \frac{dw_j}{dt} = [w_\infty(c_J) - w_J]/\tau.$$

Figure 7.5 shows a propagating Ca^{2+} wave that results when the cytosol, modeled using (21), is excited with the same initial

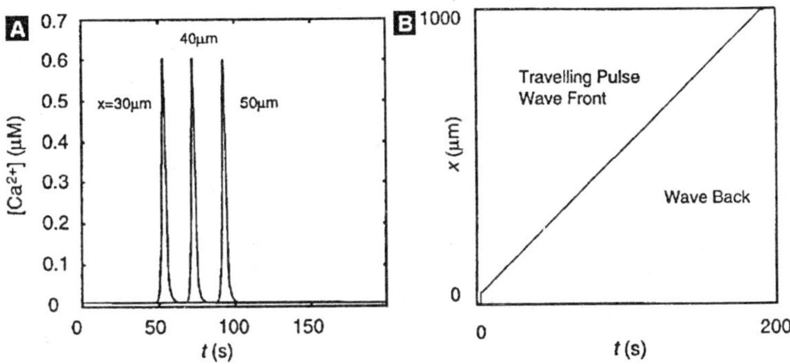

Fig. 7.5. $[Ca^{2+}]_i$ time courses and array plot similar to figure 7.2 except that parameters are changed so that the (w, $[Ca^{2+}]_i$ phase plane is excitable as in (B) of figure 7.5.

conditions as in Figure 7.2. In this case, $[Ca^{2+}]_i$ returns to basal as the Ca^{2+} transient propagates past a given point, and the cellular response is referred to as a *traveling pulse*. Traveling pulses produced by excitable dynamics are often called "trigger waves," because a perturbation at one end of the excitable medium triggers a signal that may propagate with undiminished amplitude.

Simulation of a Kinematic Wave

In previous sections we have discussed how the cytoplasm's physiological state influences Ca^{2+} wave propagation. When the cytoplasm is bistable, a likely wave phenomenon is a traveling front; conversely, an excitable cytoplasm will likely support trigger waves, *i.e.*, traveling pulses. These observations suggest that we consider the case of an oscillatory cytoplasm. IPs-mediated whole-cell Ca^{2+} responses are oscillatory in many cell types. Figure 7.6 demonstrates that it is an easy matter to adjust IP_3R parameters (Ca^{2+}-dependence of activation and inactivation) in our model so that the (*w*, $[Ca^{2+}]_i$) phase plane is oscillatory and a large-amplitude (0.01 μ*M* < $[Ca^2+]_i$ < 0.65 μM) stable limit cycle exists.

Even in the absence of diffusion an oscillatory medium will support *kinematic waves*. Figure 7.7 illustrates this possibility using an array of clocks. Depending on initial conditions (*i.e.*, the time to which the clocks are initially set), the absence of physical coupling between the clocks does not preclude the appearance of a traveling wave. Regardless of initial condition, all the clocks will eventually return to their initial state. Thus, kinematic waves

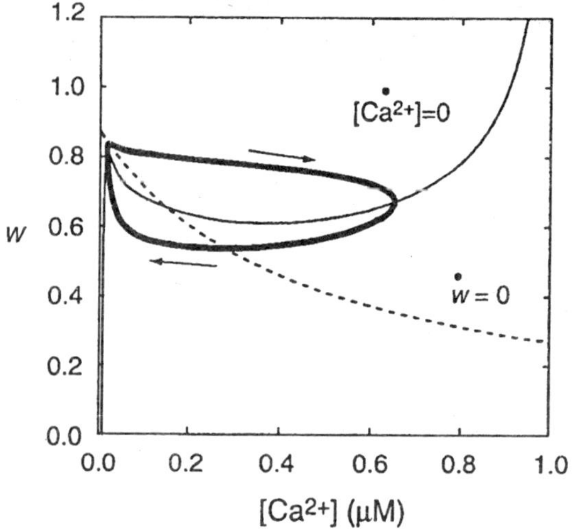

Fig. 7.6. $(w, [Ca^{2+}]_i)$ phase plane when the physiological state of the cytoplasm is oscillatory.

are distinct from traveling fronts and pulses in that they can occur with arbitrary shape.

Although a kinematic wave can be of arbitrary.shape, deceptively shaped waves may look like traveling waves when there is a certain relationship between wavelength and the oscillatory frequency of the medium. To see this, let $c(x, 0) = c_0(x)$ and $w(x, 0) = w_0(x)$ be the initial Ca^{2+} profile for our model tuned to exhibit limit cycle oscillations with period τ. If we choose $(c_0(x), w_0(x))$ "on" the limit cycle (for all x), then in the absence of Ca^{2+} diffusion the $[Ca^{2+}]_i$ will be temporally periodic at every spatial

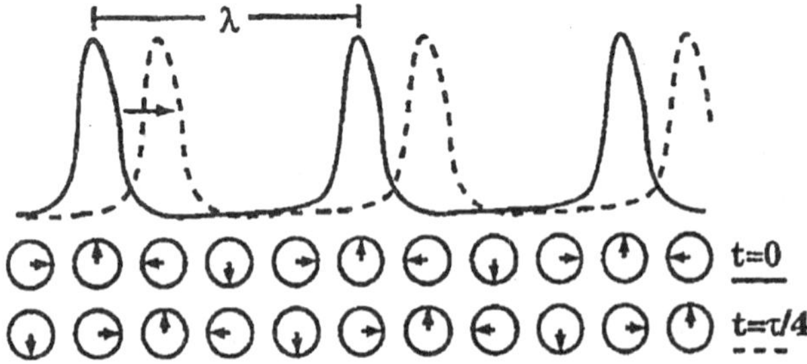

Fig. 7.7. An array of clocks illustrates a kinematic wave. The peak of the wave coincides with the hand of the clock pointing to 12 o'clock. The temporal period (τ) is the time it takes for the hand at a given location to make the complete clockwise revolution. The minimum distance between clocks with the same phase is the spatial period, or wavelength (λ).

location, *i.e.*, $c(x, t + \tau) = c(x, t)$. If the initial condition also happens to be spatially periodic, so that $c_0(x + \lambda) = c_0(x)$ (and similarly for w_0), then $[Ca^{2+}]_i$ will remain spatially periodic, *i.e.*, $c(x + \lambda, t) - c(x, t)$, where λ is the spatial period (the wavelength). As illustrated using the clock analogy in Figure 7.7, this simultaneous spatial and temporal periodicity implies a solution of the form $c(z) = c(x + vt)$ if $v = \lambda/\tau$.

Using the parameters that generated the limit cycle shown in Figure 7.6, Figure 7.8 demonstrates that oscillatory IP_3-mediated Ca^{2+} release may lead to kinematic waves. Initial conditions are chosen so that the phase of the oscillation is uniform throughout the medium, except for a central region that is phase advanced. Because Ca^{2+} diffusion is included in these simulations following (21), the distribution of phases evolves as a function of time. Because diffusion is now coupling cytoplasmic oscillations, pure kinematic waves no longer exist. Nevertheless, at each spatial location the solution remains "near" the limit cycle.

Although beyond the scope of this chapter, it can be shown that in the limit of slow diffusion (as expected here due to Ca^{2+} buffers), the relative phases of the oscillations, referred to as the *phase gradient*, evolve according to Burgers equation. The important conceptual point is that buffered Ca^{2+} diffusion causes the phase of the oscillations to synchronize over time. Figure 7.8A illustrates this with time courses $[Ca^{2+}]_i$ (x_i, t) from two spatial positions. Notice that the oscillations at these two locations synchronize; the (apparent) velocity of the repetitive wave

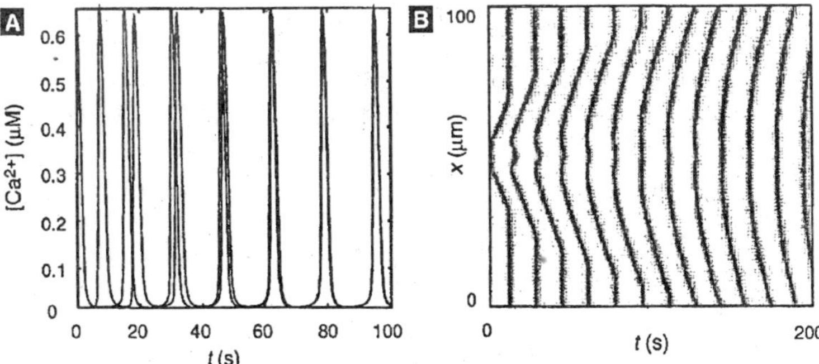

Fig. 7.8. (A) Time courses $c(x_i, t)$ at $x_i = 50$ and $63\mu m$ show that diffusion causes oscillations at different regions to synchronize. (B) An array plot illustrates the smoothing of the relative phases of the oscillations.

between these two points increases. Indeed, the velocity of a kinematic wave can be arbitrarily large because it is not a propagating signal. In the array plot of Figure 7.8B, the diminution of the phase gradient with each cycle is striking.

As discussed at length in previous sections, the presence of endogenous and exogenous Ca^{2+} buffers slows Ca^{2+} diffusion. If the physiological state of a cell's cytoplasm is oscillatory, observed wave phenomena may very well be kinematic in nature, and the influence of Ca^{2+} diffusion on propagation of the wave may be primarily through the evolution of the phase gradient. Conversely, an excitable cytbsol absolutely requires diffusion to support a propagating fronts (*i.e.*, trigger waves). While kinematic waves can exist even in the absence of diffusion, the wave speed of a propagating front scales with the effective diffusion coefficient in a predictable manner (at least in the limit of low affinity Ca^{2+} buffers). These distinctions are important and one may easily be deceived. For example, the observation of repetitive waves (*i.e.*, a *wave train*) may suggest an oscillatory medium. On the other hand, such a wave train may be due to periodic forcing (or recurrent entry) of an excitable medium, as in target patterns and spiral waves imaged in the immature *Xenopus* oocyte.

Spark-Mediated Calcium Waves

Throughout this discussion we have, without exception, made the so-called *continuum approximation*. That is, we have made no attempt to explicitly model the stochastic dynamics of individual Ca^{2+} ions undergoing Brownian motion. Indeed, an averaging process is implicit in any continuum description of matter. Consider that small molecules (such as free Ca^{2+} ions) in micromolar solution are essentially point particles with mean intermolecular spacing of approximately 100 nm; that is, in a volume of $(100\ nm)^3$ a single solute molecule will be found on average. On a distance scale of 10 nm (a factor of ten smaller) the medium is highly heterogeneous. If thermal motion of both solvent and solute could be stopped, only one out of a thousand volume elements with dimension $(10\ nm)^3$ would be found to contain a solute molecule. Clearly, at distance scales of 10 nm (or less) a continuum description is invalid.

On the other hand, if a fluorescent molecule at micromolar concentrations is optically detected using a microscope with micron resolution, the relevant volume ($1\ \mu m^3 = 10^9\ nm^3$) is likely to contain a thousand molecules. Though the actual number of molecules in the volume will vary as a function of time due to diffusion, these density fluctuations will be small. Thus, on a distance scale of a micron the medium appears spatially homogeneous, and the continuum approximation is legitimate.

The phrase "continuum approximation" can also refer to the fact that in our spatial whole-cell models we are assuming that the ER can be represented as a compartment continuously distributed throughout the cytosol, albeit with prescribed volume fraction given by f_{ER} in (8). Although it can sometimes be rigorously justified, anyone familiar with the mathematics of diffusion in heterogeneous media will recognize that this assumption is made largely for convenience. Intracellular heterogeneities abound, and only sometimes can these heterogeneities be "averaged out" in a way that justifies a bidomain description (*i.e.,* a mathematical technique known as homogenization).

One important cellular inhomogeneity known to have a profound impact on propagating IPs-mediated Ca^{2+} waves is the distribution of IP_3Rs. In the immature *Xenopus* oocyte, for example, IP_3Rs occur in clusters of 10-100 with intercluster spacing on the order of a few microns. Under some conditions this organization of Ca^{2+} release sites reveals itself in cauliflower-like wave fronts. Localized Ca^{2+} elevations due to the activation of a single Ca^{2+} release site (a Ca^{2+} puff) and even a single channel (a Ca^{2+} blip) are now regularly observed.

Localized Ca^{2+} elevations due to intracellular Ca^{2+} release have been observed in other cell types as well. When a Ca^{2+} release event is mediated by ryanodine receptors (RyRs), as in cardiac myocytes, the phenomenon is referred to as a Ca^{2+} spark. Spark-mediated waves in cardiac myocytes are amenable to modeling because Ca^{2+} waves often propagate along the longitudinal axis of a myocyte, *i.e.,* perpendicular to the orientation of the sarcomeric Z-lines. The Z-lines, transverse tubules, and Ca^{2+} release sites are oriented in the transverse direction and are regularly spaced with a separation of 2 μm. The ultrastructure of cardiac myocytes thus lends itself to one-dimensional modeling in which the cellular heterogeneity along the longitudinal axis is represented.

One such modeling approach is to augment the continuum bidomain description (8) with a spatially periodic maximum conductance for Ca^{2+} release, say, $v_{RyR}(x)$ in analogy to v_{IP_3R} in the IP_3-mediated wave model. Numerical studies show that spark-mediated Ca^{2+} waves propagate in a saltatory manner, similar to experiment. In the following section we will consider a related and even more idealized model of spark-mediated Ca^{2+} waves, the fire-diffuse-fire model.

Fire-Diffuse-Fire Model

The essence of minimal models of spark-mediated Ca^{2+} waves is to model Ca^{2+} release sites as idealized point sources, that is, an array of Dirac delta functions, denoted by $\delta(x - x_i)$, where x_i is the spatial position of the *i*th release site. While the Dirac delta function is actually not a function at all (rather, it is a distribution, the limit of a sequence of functions), it is often thought of as a sharply peaked function that is zero everywhere except $x = x_i$. In addition, the delta function is normalized so that

$$\int_{-\infty}^{\infty} \delta(x - x_i)dx = 1, \quad \text{...(22)}$$

and it has the so-called sifting property

$$\int_{-\infty}^{\infty} f(x)\delta(x - x_i)dx = f(x_i). \quad \text{...(23)}$$

Armed with delta functions, we can model spark-mediated Ca^{2+} waves by modifying the propagating front model (15) so that Ca^{2+} release occurs only at regularly spaced release sites, that is,

$$\frac{\partial c}{\partial t} = D\frac{\partial^2 c}{\partial x^2} + f(c)\sum_i \delta(x - x_i), \quad \text{...(24)}$$

where x_i is the location of the *ith* release site with kinetics given by $f(c)$.

We can further idealize this model for spark-mediated waves by removing the Ca^{2+} dependence of the release rate $f(c)$. Instead, assume that the source strength $f_i(t)$ of each release site is a square pulse function of time,

$$f_i(t) = \frac{\sigma}{\tau_R} H(t - t_i)H(t_i + \tau_R - t).$$

In this expression $H(t)$ is the Heaviside step function, so that $H(t) = 0$ for $t < 0$, $H(t) = 1$ for $t \geq 0$. The product of these Heaviside functions represents the source turning "on" at time $t = t_i$ and remaining on for duration τ_R, *i.e.*, it turns "off" at time $t = t_i + \tau_R$. The constant σ represents the source amplitude and has units of μM. μm because from (22) we see that the delta function has units of 1/μm. The normalization factor ($1/\tau_R$) is chosen so that

$$\int_{-\infty}^{\infty} f_i(t)\, dt = \sigma \qquad ...(25)$$

Substituting this form for the Ca^{2+} release rate into (24) gives the "fire-diffuse-fire" model of spark-mediated Ca^{2+} wave propagation,

$$\frac{\partial c}{\partial t} = D\frac{\partial^2 c}{\partial x^2} + \frac{\sigma}{\tau_R}\sum_i \delta(x - x_i)H(t - t_i)H(t_i + \tau_R - t). \qquad ...(26)$$

To finish our presentation of the fire-diffuse-fire model, we must specify the location of each release site (x_i) and the times (t_i) at which Ca^{2+} release begins. If we further assume a regular array of release sites, then $x_i = id$ where d is the site spacing. Because we have Ca^{2+}-induced Ca^{2+} release in mind, we assume that Ca^{2+} release at site i begins when the local Ca^{2+} concentration $c(x_i, t)$ achieves a fixed threshold Ca^{2+} concentration c_θ (here 0.1 μM).

Interestingly, the fire-diffuse-fire model supports continuous as well as propagating Ca^{2+} signals. Here spark-like Ca^{2+} releases lead to a propagating signal that is distinctly saltatory. Note that the continuous wave is traveling at 11.3 μm/s. while the saltatory wave is traveling at 67 μm/s. The long duration of Ca^{2+} release in the continuous case appears to slow the velocity of the propagating signal.

The fire-diffuse-fire model can be analyzed to give insight into the continuous and saltatory limits of Ca^{2+} wave propagation. As the reader may expect, it is not τ_R alone but rather a dimensionless parameter that determines the existence and form of propagating signals. Indeed, it can be shown that the relevant dimensionless quantity is $D\tau_R/d^2$. The continuous limit corresponds to $D\tau_R/d^2 \gg 1$ and the saltatory limit to $D\tau_R/d^2 \ll 1$. Below we study these two limits separately to determine (you guessed it) how diffusion influences wave velocity in both limits.

Analysis of the Continuous Limit

The continuous limit pertains when $D\tau_R/d^2 \gg 1$, that is, when diffusion is fast and the release time long compared to the intersite separation. This limit could be achieved in a simulation by increasing the density of release sites ($d \to 0$) while simultaneously decreasing the release rate (σ) so that the release per unit length (σ/d) is constant.

Because the thresholding that determines Ca^{2+} release times (t_i) is the only nonlinear part of our model (5.26), we can convolve the release rates $f_i(t)$ with the diffusion kernel to write an explicit expression for $c(x, t)$:

$$c(x, t) = \frac{\sigma}{\tau_R} \sum_{i=-(N-1)}^{N-1} \int_{t_i}^{\min(t, t_i+\tau_R)} \frac{dt'}{\sqrt{4\pi D(t-t')}} \exp\left[-\frac{(x-x_i)^2}{4D(t-t')}\right], \text{ for } t > t_{N-1}, \quad ...(27)$$

where we have assumed a symmetric profile $c(x, t)$ and the firing of $N - 1$ sites on either side of the origin. Equation (27) is not yet a solution because we have the unknown parameters t_i to determine. To find the time at which site N fires we evaluate (27) at $(t = t_N, x = x_N)$ and set $c(x_N, x_N) = c_\theta$. This yields

$$c_\theta = \frac{\sigma}{\tau_R} \sum_{i=-(N-1)}^{N-1} \int_{t_i}^{\min(t, t_i+\tau_R)} \frac{dt'}{\sqrt{4\pi D(t-t')}} \exp\left[-\frac{(x-x_i)^2}{4D(t-t')}\right],$$

an expression that can be solved for t_N. At this point we seek traveling-wave like solutions with regular firing times. This implies that $t_i = i\Delta$, where Δ is the time interval between adjacent site firings and a velocity of propagation given by $v = d/\Delta$. By substituting $\Delta = t_N/N$ and solving for Δ it can be shown that the

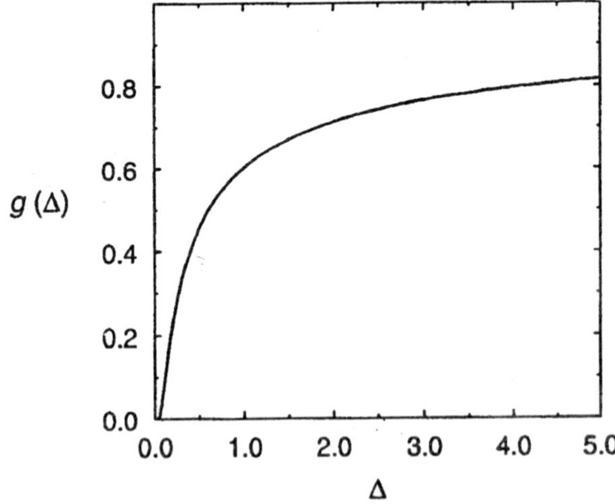

Fig. 7.9. Relationship between $\alpha = c_\theta d/\sigma$ and the dimensionless firing interval ($\Delta = 1/v$) for the fire-diffuse-fire model in the saltatory limit.

velocity of such a propagating wave has the following dependence on model parameters:

$$v \approx \sqrt{\left(\frac{\sigma}{dc_\theta}\right)\left(\frac{D}{\tau_R}\right)}. \quad ...(28)$$

Notice that the factor of $\sqrt{D/\tau_R}$ is similar to what we have seen before for traveling fronts and pulses. The existence of traveling wave solutions depends on $\sqrt{\sigma/dc_\theta}$.

Analysis of the Saltatory Limit

The saltatory limit of the fire-diffuse-fire model corresponds to $D\tau_R/d^2 \ll 1$ in (26), that is, when diffusion is slow and release time short compared to the intersite separation. This limit could be achieved in a simulation by decreasing the time constant for Ca^{2+} release (τ_R) while maintaining a fixed density of release sites (d constant). Because the normalization factor ($1/\tau_R$) is chosen to satisfy (25) regardless of τ_R, the release rate per unit length (σ/d) remains constant as $\tau_R \to 0$, and (26) becomes

$$\frac{\partial c}{\partial t} = D\frac{\partial^2 c}{\partial x^2} + \sigma \sum_i \delta(x - x_i)\,\delta(t - t_i). \quad ...(29)$$

We now nondimensionalize space ($x^* = x/d$, $x_i^* = x_i/d = i$), time ($t^* = tD/d^2$, $t_i^* = t_iD/d^2$), and concentration ($c^* = c/c_\theta$) and drop asterisks to write

$$\frac{\partial c}{\partial t} = \frac{\partial^2 c}{\partial x^2} + \frac{1}{\alpha}\sum_i \delta(x - x_i)\,\delta(t - t_i), \quad ...(30)$$

where $a = c_\theta d/\sigma$. Using the diffusion kernel we obtain an implicit expression for $c(x, t)$,

$$x(x, t) = \sum_i H(t - t_i)\sqrt{\frac{1}{4\pi\alpha^2(t - t_i)}}\exp\left[-\frac{(x - i)^2}{4(t - t_i)}\right], \quad ...(31)$$

where the presence of the Heaviside function indicates no contribution from sites that have not yet fired.

Assuming that the sites for which $-N < i < N$ have fired at times t_i through t_N (left and right pairs simultaneously) we want to determine t_{N+1}. The dimensionless threshold for firing is now $c = 1$, so from (31) we have

$$c(x_{N+1}, t_{N+1}) = 1 = \sum_{i=-N}^{N}\sqrt{\frac{1}{4\pi\alpha^2(t_{N+1} - t_i)}}\exp\left[\frac{(N + 1 - i)^2}{4(t_{N+1} - t_i)}\right]. \quad ...(32)$$

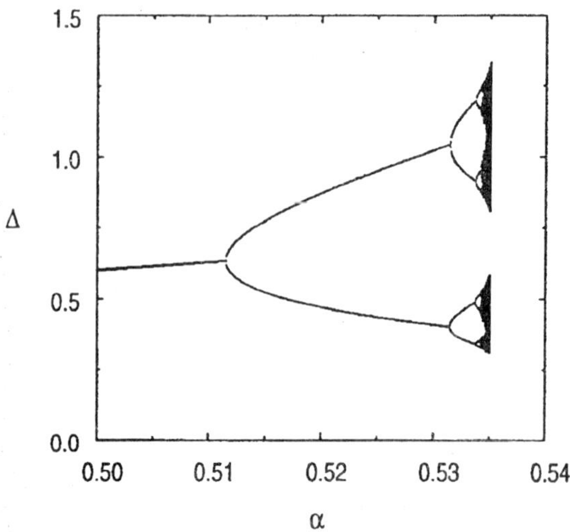

Fig. 7.10. For a range of values for α, the saltatory limit of the fire-diffuse-fire model is simulated by successively calculating the dimensionless firing interval (Δ) for n = 1, 2, ... using (eq. 32) and the criterion c(n, t_i) = 1.

Because we are interested only in long time solutions, we consider the large N limit and neglect all the terms in the above sum with $i \le 0$, that is, we are following a rightward traveling wave that eventually is not influenced by the sites to the left of the origin. Under this approximation (32) simplifies to

$$\alpha = \sum_{i=1}^{N} \sqrt{\frac{1}{4\pi(t_{N+1} - t_i)}} \exp\left[\frac{(N+1-i)^2}{4(t_{N+1} - t_i)}\right]. \qquad ...(33)$$

This expression is an implicit map for t_{N+1} as a function of all the previous firing times, $(t_N, t_{N-1}, ... t_1) \rightarrow t_{N+1}$. traveling-wave-like solutions correspond to fixed points of this map with regular firings, that is, $t_{N+1} = t_{N+\Delta}$ with constant Δ (giving a dimensionless velocity of $v = 1/\Delta$). Substituting $\Delta = t_{N+1} - t_N$ in (33) we obtain

$$\alpha = \sum_{i=1}^{N} \sqrt{\frac{1}{4\pi\Delta\,(N+1-i)}} \exp\left[\frac{(N+1-i)}{4\Delta}\right]. \qquad ...(34)$$

Defining $n = N + 1 - i$ and taking the limit $N \rightarrow \infty$ we obtain

$$\alpha = \sum_{n=1}^{\infty} \sqrt{\frac{1}{4\pi n\Delta}} \exp(-n/4\Delta) \equiv g(\Delta). \qquad ...(35)$$

It can be shown that $0 \le g(\Delta) \le 1$. The first equality holds in the high-velocity limit ($\Delta \rightarrow 0$) and the second for low velocity (Δ

$\to \infty$). Since $g(\Delta)$ is monotonic, we can numerically calculate a unique solution $\Delta = g^{-1}(\alpha)$ when $0 < \alpha < 1$, *i.e.*, the range of $g(\Delta)$. At this point it is worth remembering that $\alpha = c_\theta d/\sigma$. Thus, a necessary condition for the existence of a velocity is that $\alpha < 1$. If the sites are too far apart or too weak or the threshold too high, there can be no propagating waves. Interestingly, it can be shown that propagation failure occurs through a sequence of instabilities as a is increased well before the condition $\alpha = 1$ is reached (see further reading). When a is small ($\sigma/dc_\theta \gg 1$), propagating wave solutions exist and are stable. It can be shown that the dimensional velocity v in this saltatory propagation limit is given by

$$v \approx \frac{4D}{d}\ln\left(\frac{\sigma}{dc_\theta}\right); \qquad ...(36)$$

that is, it scales linearly with D, quite unlike propagating waves in the continuous limit, which have velocity that scales with $\sqrt{D}$ as in (28). If fact, whenever τ_R is sufficiently small (even if α is not particularly small), the saltatory wave velocity predicted by the fire-diffuse-fire model scales linearly with D.

Modeling Localized Calcium Elevations

Our discussion of spark-mediated Ca^{2+} wave propagation began with a discussion of cellular heterogeneity. During this discussion of Ca^{2+} wave phenomena it is important to remember that many cellular processes (including synaptic transmission, activity-dependent synaptic plasticity, and regulation of neuronal excitability) can be initiated by changes in intracellular Ca^{2+} concentration in the absence of a global response such as a Ca^{2+} oscillation or wave. For this reason, Ca^{2+} sparks, puffs, and localized Ca^{2+} elevations near voltage gated plasma membrane Ca^{2+} channels (sometimes called Ca^{2+} microdomains) are cellular signals of great interest. Localized Ca^{2+} elevations are not only the "building blocks" of global Ca^{2+} release events, but also highly specific regulators of cellular function.

As discussed above in the context of experimental observations of Ca^{2+} waves, the interpretation of microfluorometric measurements of Ca^{2+} puffs and sparks is complicated by the diffusion of Ca^{2+}, endogenous buffers, and indicator, all of which contribute to the dynamics of a fluorescence signal during and after Ca^{2+} release. While in the case of global Ca^{2+} responses and Ca^{2+} waves the equilibrium relation will likely hold between Ca^{2+} and indicator, this is not so easily demonstrated in the case of localized Ca^{2+} elevations. A further complicating factor in interpreting *confocal* microfluorometric measurements is the optical blurring that occurs due to limited spatial resolution.

The reaction-diffusion equations for the buffered diffusion of intracellular Ca^{2+}, (10)-(12), are t' e starting point of a theoretical understanding the dynamics of localized Ca^{2+} elevations. In the simplest scenario, a Ca^{2+} puff or spark is due to Ca^{2+} release through one channel or a tight cluster of channels. If Ca^{2+} is released from intracellular Ca^{2+} stores deep within a large cell (so that the plasma membrane is far away and doesn't influence the time-course of the event), and the intracellular milieu is homogeneous and isotropic, then we have spherical symmetry. In this case, the evolving profiles of Ca^{2+} and buffer (though a function of time and distance from the source) will not be functions of the polar (ϕ) or azimuthal (θ) angle. In the case of such spherical or radial symmetry the Laplacian reduces to

$$\nabla^2 = \frac{1}{r^2}\frac{\partial}{\partial r}\left[r^2 \frac{\partial}{\partial r}\right] = \frac{\partial^2}{\partial r} + \frac{2}{r}\frac{\partial}{\partial r}. \qquad ...(37)$$

Figure 7.11 shows a spherically symmetric calculation of a localized Ca^{2+} elevation using the full equations for the buffered diffusion of Ca^{2+}, (10)-(12), with parameters consistent with measurements of the effective diffusion coefficient in *Xenopus* oocyte cytoplasm. Figure 7.11 is a numerically calculated snapshot of the concentration profiles for each species after an elapsed time of 10 ms. The concentration $[Ca^{2+}]_i$ is elevated near the source (*thick solid line*). Because released free Ca^{2+} reacts with buffer, the concentration of bound buffer *(thin solid and thick dashed lines)* is elevated near the source. Conversely, the concentration of free buffer *(thin dotted and dashed lines)* decreases near the source. In this simulation, 250 μM stationary buffer was included in addition to 50 μM mobile buffer (both with K of 10 μM). A source amplitude of 5 pA was used, corresponding to a tight cluster of IP_3Rs. Interestingly, Figure 7.11 shows that the mobile buffer is less easily saturated than stationary buffer, in spite of the fact that the stationary buffer is at fivefold higher concentration.

Simulations such as these have played a role in understanding the dynamics of puffs and sparks. Figure 7.12 shows a Ca^{2+} spark simulated using parameters consistent with

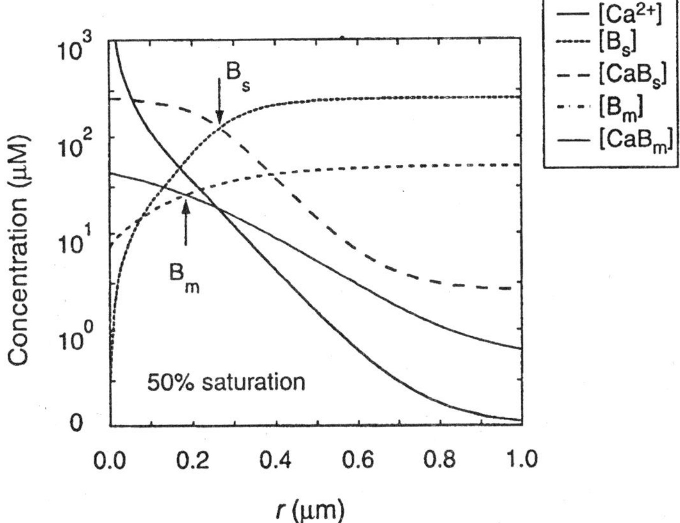

Fig. 7.11. Representative calculation of Ca^{2+} profile near a point source for free Ca^{2+}. Source amplitude (σ) and elapsed time are 5 pA and 10 ms respectively.

experimental observation in cardiac myocytes. Such simulations confirm that the timecourse of observed fluorescence can be explained by a 2 pA, 15 ms Ca^{2+} -release event from a tight cluster of RyRs located on the sarcoplasmic reticulurn membrane.

Parameter studies using this model indicate that Ca^{2+} spark properties (such as brightness, full width at half maximum, and decay time constant) are very dependent on indicator dye parameters (such as association rate constant, concentration, and diffusion coefficient). These relationships are not always intuitive. For example, increasing indicator dye concentration decreases the brightness of the simulated Ca^{2+} spark in Figure 7.12. This is partly due to the fact that spark brightness is a normalized measure (peak/basal fluorescence), and partly due to the fact that high concentrations of indicator perturb the underlying free Ca^{2+} signal.

Steady-State Localized Calcium Elevations

Numerical simulations like those in Figure 7.11 confirm that localized Ca^{2+} elevations achieve steady-state values very rapidly (within microseconds) near point sources. Steady-state solutions to the full equations are thus of interest because they allow estimates of "domain" Ca^{2+} concentration near open Ca^{2+} channels.

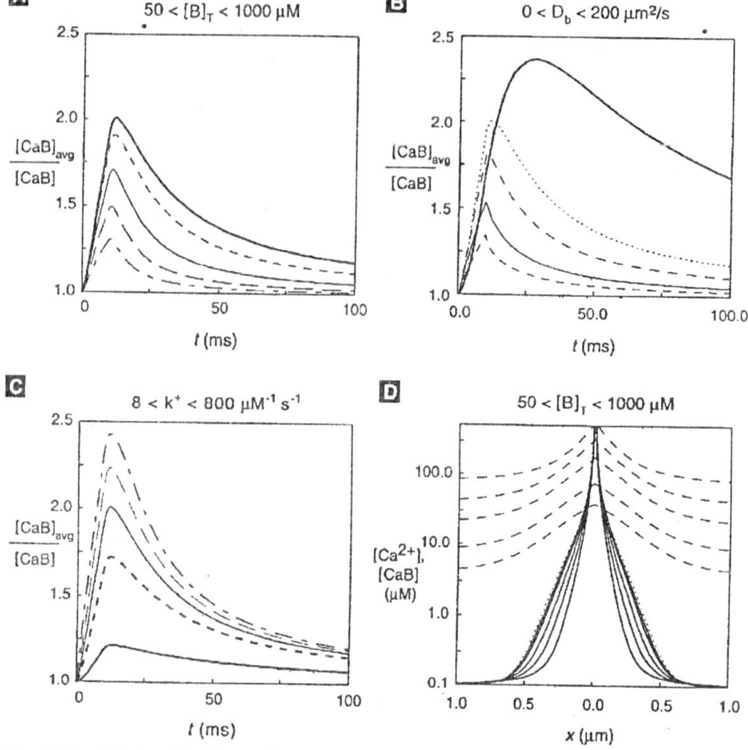

Fig. 7.12. Effects of indicator dye parameters on Ca^{2+} spark properties. (A,B,C) Time course of normalized blurred fluorescence signal estimated according to (eq. 2) with $\eta_B = 0$. (D) Snapshot of $[Ca^{2+}]_i$ and [CaB] profiles before terminaion of Ca^{2+} release.

These steady-state solutions lend themselves to analysis, giving insight into the limiting (long time) "shape" of localized Ca^{2+} elevations. In the case of one mobile buffer, steady-state solutions to (10)-(12) will satisfy the following boundary value problem:

$$0 = D_C \nabla^2 [\mathrm{Ca}^{2+}]_i - k^+ [\mathrm{B}][\mathrm{Ca}^{2+}]_i + k^- ([\mathrm{B}]_\mathrm{T} - [\mathrm{B}]), \quad \text{...(38)}$$

$$0 = D_B \nabla^2 [\mathrm{B}] - k^+ [\mathrm{B}][\mathrm{Ca}^{2+}]_i + k^- ([\mathrm{B}]_\mathrm{T} - [\mathrm{B}]), \quad \text{...(39)}$$

with associated boundary conditions

$$\lim_{r\to 0} \left\{ -4\pi r^2 D_C \frac{d[\mathrm{Ca}^{2+}]_i}{dr} \right\} = \sigma, \qquad \lim_{r\to\infty} [\mathrm{Ca}^{2+}]_i = [\mathrm{Ca}^{2+}]_\infty,$$

$$\lim_{r\to 0} \left\{ -4\pi r^2 D_B \frac{d[\mathrm{B}]}{dr} \right\} = 0, \qquad \lim_{r\to\infty} [\mathrm{B}] = [\mathrm{B}]_\infty = \frac{K[\mathrm{B}]_\mathrm{T}}{K + [\mathrm{Ca}^{2+}]_\infty}. \quad \text{...(40)}$$

Here we have eliminated the equation for [CaB] (rather than [B] as before) and written D_C and D_B for the diffusion coefficients of free Ca^{2}+and free buffer, respectively.

Fixed buffers, while important for the time-dependent evolution of localized Ca^{2+} elevations, have no influence on steady states. This can be seen by inspecting (10)-(12), where $D_{CB} = 0$ implies $R = -k_+[B][Ca^{2+}]_i + k_-[CaB] = 0$ at steady state.

As discussed previously, there are many advantages to non-dimensionalizing equations before preceding to analyze them. A convenient nondi-mensionalization of (38)-(40) begins by scaling $[Ca^{2+}]_i$ and [B] by representative concentrations, the dissociation constant (K) and total concentration of buffer ($[B]_T$). This gives two dimensionless dependent variables, $\hat{c}$ and $\hat{b}$, given by $\hat{c} = [Ca^{2+}]_i/K$ and $\hat{b} = [B]/[B]_T$.

The reader can confirm that nondimensionalizing the independent variable ($\hat{\rho} = r/L$ with $L = \sigma/4\pi D_C K$) simplifies (38) and (39) to

$$\hat{\varepsilon}_c \nabla^2_{\hat{\rho}} \hat{c} - (\hat{c}\hat{b} + \hat{b} - 1) = 0, \qquad ...(41)$$

$$\hat{\varepsilon}_b \nabla^2_{\hat{\rho}} \hat{b} - (\hat{c}\hat{b} + \hat{b} - 1) = 0, \qquad ...(42)$$

where the subscript on the Laplacian indicates that differentiation is with respect to $\hat{\rho}$.

Two dimensionless diffusion coefficients ($\hat{\varepsilon}_c$ and $\hat{\varepsilon}_b$) appear in these equations. In terms of the original dimensional parameters of the problem, they are given by $\hat{\varepsilon}_c = \varepsilon\alpha$ and $\hat{\varepsilon}_b = \varepsilon D$, where $\alpha = K/[B]_T$ is a buffering factor (small when buffer is at high concentration compared to the dissociation constant), $D = D_B/D_C$ is a relative diffusion coefficient between buffer and Ca^{2+}, and the common factor e is given by

$$e = (4\pi)^2 D^3_C K/\sigma^2 k^+, \qquad ...(43)$$

a quantity that is small for strong sources and/or fast buffers.

Steady-State Excess Buffer Approximation (EBA)

The dimensionless steady-state equations for the buffered diffusion of Ca^{2+} near a point source, (41) and (42), are nonlinear, and no general analytical solution is known for these equations. However, we can begin to understand the behavior of solutions (and the effect of Ca^{2+} buffers on Ca^{2+} domains) by considering (41) and (42) in limiting parameter regimes. The first such limit we will consider is called the "excess buffer approximation." If buffer is in excess, then the parameter $\alpha = K/[B]_T$ will be very small, and $\hat{\varepsilon}_c \approx \varepsilon\alpha$ will also be small when $\varepsilon = O(1)$. Therefore, we consider in detail (41) and (42) when $\hat{\varepsilon} \approx 0$, which in physical terms implies that the diffusion coefficient of c is small compared to the size of the reaction terms in (41).

The mathematically inclined reader will notice that this is a singular perturbation problem. Because this technique goes beyond the scope of this chapter, we present only a heuristic analysis here. The interested reader is invited to consult Smith *et al.* (2001) for a more rigorous treatment.

With this caveat, we formally set $\hat{\varepsilon}_c = 0$ in (41), giving

$$\hat{c}\hat{b} + \hat{b} - 1 = 0, \qquad \text{...(44)}$$

which implies that (42) simplifies to

$$\nabla^2 \hat{b} = 0$$

When combined with the boundary conditions for b, this equation implies

$$\hat{b} = \hat{b}_\infty, \qquad \text{...(45)}$$

where $\hat{b}_\infty = 1/(1 + \hat{c}_\infty)$ is the fraction of free buffer far from the source. Thus, our assumption that buffer is in excess ($\hat{\varepsilon}_c = 0$) implies that the buffer is not perturbed from its equilibrium value $\hat{b}_\infty$. Substituting (45) into (41) gives

$$\hat{\varepsilon}_c \nabla^2 \hat{c} - \hat{b}_\infty (\hat{c} - \hat{c}_\infty) = 0.$$

This is a linear equation satisfied by

$$\hat{c} = \frac{1}{\hat{\rho}} e^{-\hat{\rho}/\Lambda} + \hat{c}_\infty, \qquad \text{...(46)}$$

where the dimensionless space constant $\Lambda = \sqrt{\hat{\varepsilon}_c / \hat{b}_\infty}$ and $\hat{c}_\infty = [Ca^{2+}]_i / K$ are chosen to satisfy the boundary conditions on $\hat{c}$ that can be derived from (40).

When this result is expressed in dimensional form, we have

$$[Ca^{2+}]_i = \frac{\sigma}{4\pi D_C r} e^{-r/\lambda} + [Ca^{2+}]_\infty, \qquad \text{...(47)}$$

where λ. is the characteristic length constant for the mobile Ca^{2+} buffer given by $\lambda = \sqrt{D_C / k^+ [B]_\infty}$. This *excess buffer approximation,* first derived by Neher, is valid when mobile buffer is in high concentration and/or when the source amplitude is small, that is, $\lim_{r \to 0} [B] \approx [B]_\infty$. Note that λ decreases with increasing association rate constant (k^+) and free buffer concentration far from the source ($[B]_\infty$). When a buffer is in excess, we thus expect further increases in concentration to restrict localized Ca^{2+} elevations. In addition, buffers with fast-reaction kinetics are expected to restrict localized Ca^{2+} elevations more than slow buffers.

Steady-State Rapid Buffer Approximation (RBA)

The *steady-state rapid buffer approximation* near a point source for Ca^{2+} can be derived by noticing in (43) that rapid buffer (large k^+) leads to small values of ε. This results in small values for both $\hat{\varepsilon}_c$ and $\hat{\varepsilon}_b$, which in physical terms implies that the diffusion coefficient of both $\hat{c}$ and $\hat{b}$ are small compared to the size of the reaction terms in (41) and (42).

If we formally set $\hat{\varepsilon}_c = \hat{\varepsilon}_b = 0$ in these equations, we find that, as before, (44) holds. Solving for $\hat{b}$, we find that at every spatial location $\hat{b}$ is given by

$$\hat{b} = \frac{1}{1+\hat{c}}, \qquad ...(48)$$

or in dimensional terms,

$$[BJ] = \frac{K[B]_T}{K+[Ca^{2+}]_i}. \qquad ...(49)$$

These equations are statements of *local equilibrium*, the fundamental assumption used in deriving the rapid buffer approximation in the context of traveling waves.

We proceed with this derivation of the steady-state RBA by subtracting (42) from (41) to give

$$\nabla^2_{\hat{\rho}}\left(\hat{\varepsilon}_c\hat{c} - \hat{\varepsilon}_b\hat{b}\right) = 0, \qquad ...(50)$$

In physical terms this expression is equivalent to the statement that at steady state the flux of total Ca^{2+}, diffusing in both free and bound forms across any spherical surface centered on the source is equal to the flux entering through the source. Integrating twice with respect to p and using the boundary conditions to determine the integration constants gives

$$\hat{\varepsilon}_c\hat{c} - \hat{\varepsilon}_b\hat{b} = \frac{\hat{\varepsilon}_c}{\hat{\rho}} + \hat{\varepsilon}_c\hat{c}_\infty - \hat{\varepsilon}_b\hat{b}_\infty. \qquad ...(51)$$

Substituting (48) into this equation gives

$$\hat{\varepsilon}_c\hat{c} - \hat{\varepsilon}_b\left(\frac{1}{1+\hat{c}}\right) = \frac{\hat{\varepsilon}_c}{\hat{\rho}} + \hat{\varepsilon}_c\hat{c}_\infty - \hat{\varepsilon}_b\hat{b}_\infty, \qquad ...(52)$$

which upon solving for $\hat{c}$ and converting back into dimensional form gives the steady-state RBA

$$[Ca^{2+}]_i = \frac{1}{2D_C}\left(-D_C K + \frac{\sigma}{4\pi r} + D_C[Ca^{2+}]_\infty - D_B[B]_\infty\right.$$

$$\left. + \sqrt{\left(D_C K + \frac{\sigma}{4\pi r} + D_C[Ca^{2+}]_\infty - D_B[B]_\infty\right)^2 + 4D_C D_B[B]_T K}\right) \quad ...(53)$$

The steady-state RBA tends to be valid when ε is small, *e.g.*, when buffers have large association and dissociation rate constants (k^+ and k^-). Interestingly, a sufficiently large source amplitude (σ) can compensate for modest binding rates, also causing ε to be small and the steady-state RBA to be valid.

Complementarity of the Steady-State EBA and RBA

The fundamental assumptions used in deriving the steady-state excess and rapid buffer approximations are significantly different. In the case of the RBA (53) we assumed that buffer and Ca^{2+} were in local equilibrium, (49). Because $[Ca^{2+}]_i \rightarrow \infty$ as $r \rightarrow 0$ in (53), we see that according to the steady-state RBA

$$\lim_{r\to 0}[B] \approx 0 \text{ (RBA).} \quad ...(54)$$

Thus, the steady-state RBA cannot be valid unless the source is strong enough to saturate the buffer. On the other hand, in our derivation of the EBA we assumed that the buffer is not perturbed from its equilibrium value, (45). If this is true even near the source, then

$$\lim_{r\to 0}[B] \approx [B]_\infty \text{ (EBA).} \quad ...(55)$$

Thus, we expect the EBA and RBA approximations to be complementary, in the sense that the steady-state solution to the full equations for the buffered diffusion of Ca^{2+} near a point source (the correct answer) cannot simultaneously be EBA-like, as in (55), and RBA-like, as in (54). In the process of extending both the EBA and RBA to higher order, this expectation has been confirmed.

8

Computational Algorithms

Many bioinformatics algorithms may be formulated in the language of *graph theory*. The use of the word "graph" here is different than in many physical science contexts: we do not mean a chart of data in a Cartesian coordinate system. In order to work with graphs, we will need to define a few concepts that may not appear at first to be particularly well motivated by biological examples, but after introducing some of the mathematical theory we will show how powerful they can be in such bioinformatics applications as DNA sequencing and protein identification.

Graphs

Figure 8.1 (a) shows two white and two black knights on 3 × 3 chessboard. Can they move using the usual chess knight's moves, to occupy the position shown in Figure 8.1 (b)? Needless to say, two knights cannot occupy the same square while they are moving.

Figure 8.2b represents the chessboard as a set of nine points. Two points are connected by a line if moving from one point to another is a valid knight move. Figure 8.2c shows an equivalent representation of the resulting diagram that reveals that knights move around a "cycle" formed by points 1, 6, 7, 2, 9, 4, 3 and 8. Every knight's move on the chessboard corresponds to moving to a neighboring point in the diagram, in either a clockwise or counterclockwise direction. Therefore the white-white-black black knight arrangement cannot be transformed into the alternating white-black-white-black arrangement.

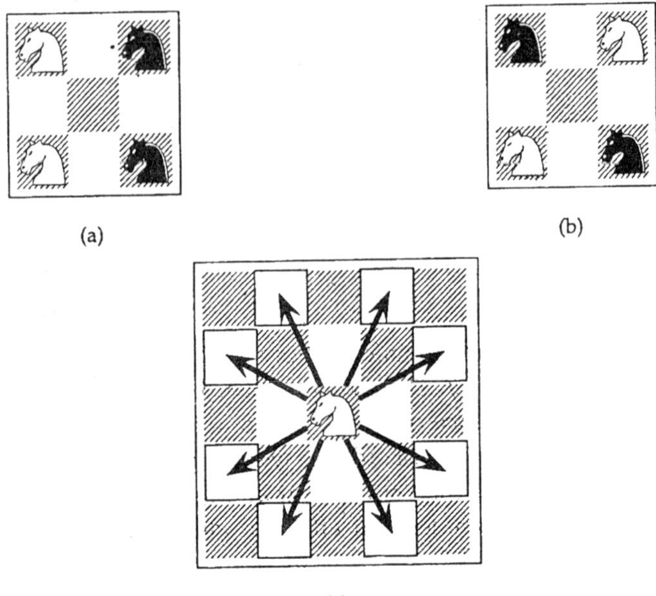

Fig. 8.1. Two configurations of four knights on a chessboard. Can you have valid knight moves to turn the configuration in (a) into the configuration in (b)? Valid knight moves are shown in (c).

Diagrams with collections of points connected by lines are examples of *graphs*. The points are called *vertices* and lines are called *edges*. A simple graph shown in Figure 8.3, consists of five vertices and six edges. We denote a graph by $G = G(V, E)$ and describe it by its set of vertices V and set of edges E (every edge an be written as a pair of vertices). The graph in figure 8.3 is described by the vertex set $V = \{a, b, c, d, e\}$ and the edge set $E = \{(a, b), (a, c) (b, c), (b, d), (c, d), (c, e)\}$.

The way the graph is actually drawn in irrelevant; two graphs with the same vertex and edge sets are equivalent, even if the particular pictures that represent the graph appear different. The only important feature of a graph is which vertices are connected and which are not.

Figure 8.4 represents another chessboard obtained from a 4×4 chessboard by removing the four corner squares. Can a knight travel around this board, pass through each square exactly once, and return to the same square it started on? Figure 8.4 (b) shows a rather complex graph with twelve vertices and sixteen

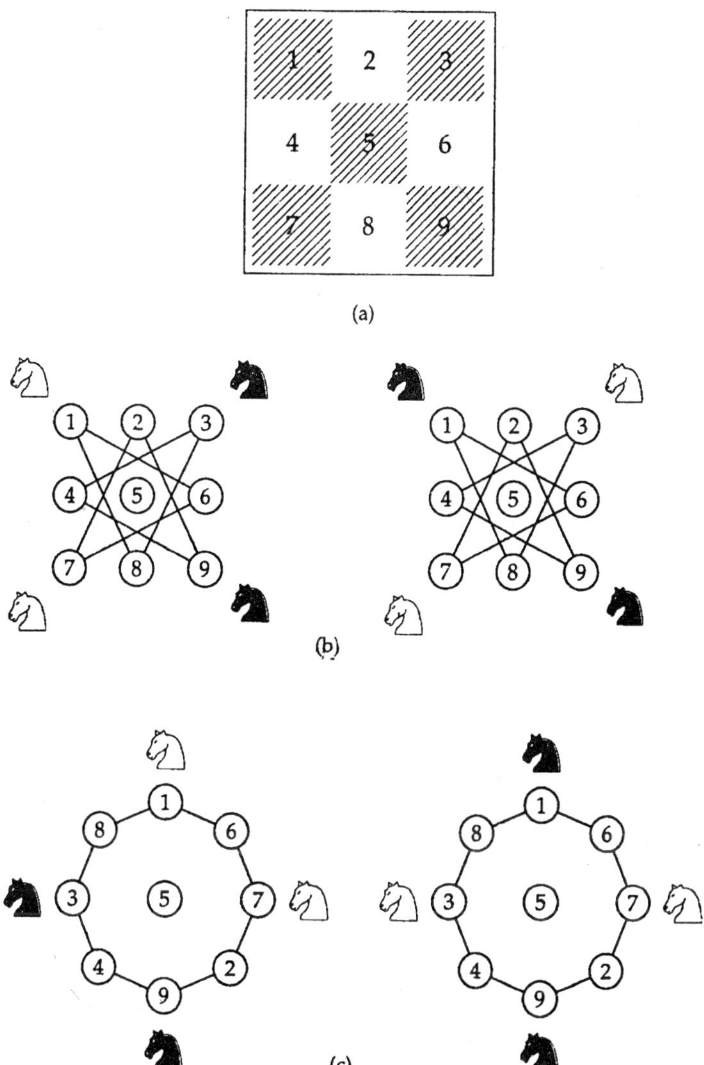

Fig. 8.2. A graph representation of a chessboard. A knight sitting on some square can reach any of the squares attached to that square by an edge.

edges revealing all possible knight moves. However, rearranging the vertices reveals the cycle that describes the correct sequence of moves. The number of edges incident to a given vertex v is called the *degree* of the vertex and is denoted $d(v)$. For example, vertex 2 in figure 8.4 **(c)** has degree 3 while vertex 4 has degree 2.

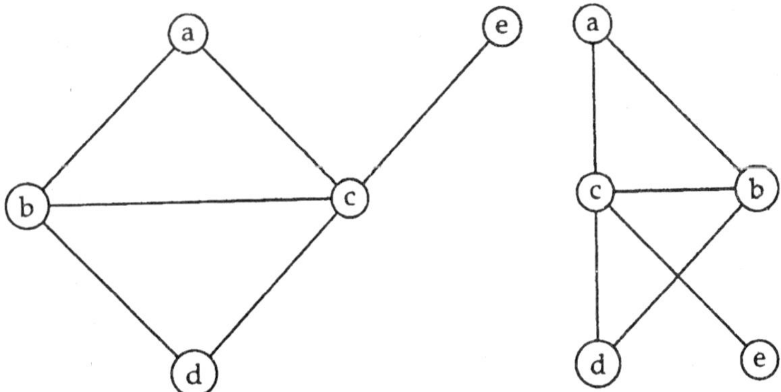

Fig. 8.3. Two equivalent representations of a simple graph with five vertices and six edges.

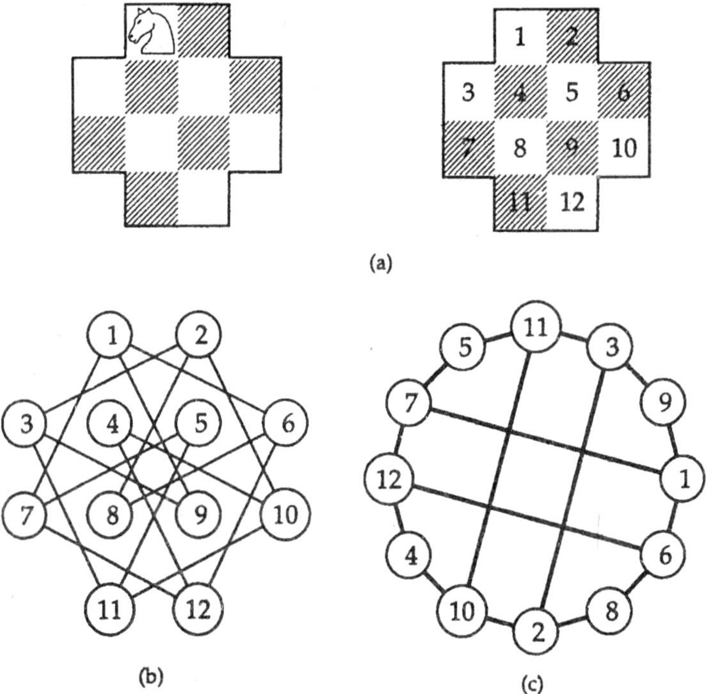

Fig. 8.4. The knight's tour through the twelve squares in part (a) can be seen by constructing a graph (b) and rearranging its vertices in a cleave way (c).

The sum of degrees of all 12 vertices is, in this case, 32 (8 vertices of degree 3 and 4 vertices of degree 2), twice the number of edges in the graph. This is not a coincidence: for every graph G with vertex set V and edge set E, $\Sigma_{v \in V} d(v) = 2 \cdot |E|$.

Indeed, an edge connecting vertices v and w is counted in the $\Sigma_{v \in V} d(v)$ twice: first in the term $d(v)$ and again in the term $d(w)$. The equality $\Sigma_{v \in V} d(v) = 2 \cdot |E|$ explains why you cannot connect fifteen phones such that each is connected to exactly seven others, and why country with exactly three roads out of every city cannot have precisely 100 roads. Many bioinformatics problems make use of *directed* graphs, in which every edge is directed from one vertex to another, as shown by the arrows in figure 8.5.

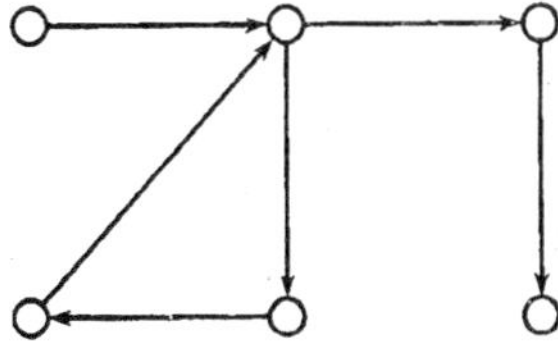

Fig. 8.5. A directed graph.

Every vertex v in a directed graph is characterized by *indegree*(v) (the number of incoming edges) and *outdegree*(v) (the number of outgoing edges). For every directed graph $G(V, E)$,

$$\sum_{v \in V} indigree(v) = \sum_{v \in V} outdegree(v),$$

since every edge is counted once on the right-hand side of the equation and once on the left hand side.

A graph is called *connected* if all pairs of vertices can be connected by a *path*, which is continuous sequence of edges, where each successive edge begins where the previous one left off. Paths that start and end at the same vertex are referred to as *cycle*. For example, the paths (3-2-10-11-3) and (3-2-8-6-12-7-5-11-3) in figure 8.4 (c) are cycles. Graphs that are not connected are *disconnected*. Disconnected graphs can be partitioned into *connected components*.

One can think of a graph as a map showing cites (vertices) and the freeways (edges) that connect them. Not all cities are connected by freeways: for example, you cannot drive from Miami to Honolulu. These two cities belong to two different connected components of the graph. A graph is called *complete* if there is an edge between every two vertices. *Graph theory* was born in eighteenth century when Leonhard Euler solved the famous Konigsberg Bridge problem. Konigsberg is located on the banks

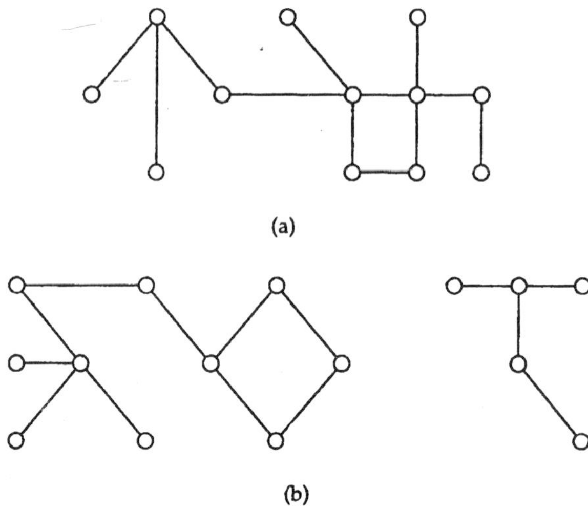

Fig. 8.6. A connected (a) and a disconnected (b) graph.

of the Pregel River, with a small island in the middle. The various parts of the city are connected by bridges and Euler was interested in whether he could arrange a tour of the city in such a way that the tour visits each bridge exactly once. For Konigsberg this turned out to be impossible, but Euler basically invented an algorithm to solve this problem for any city. Figure 8.8 shows a city map with ten islands and sixteen bridges, as well as the transformation of the map into a graph with ten vertices and sixteen edges (every island corresponds to a vertex and every bridge corresponds to an edge). After this

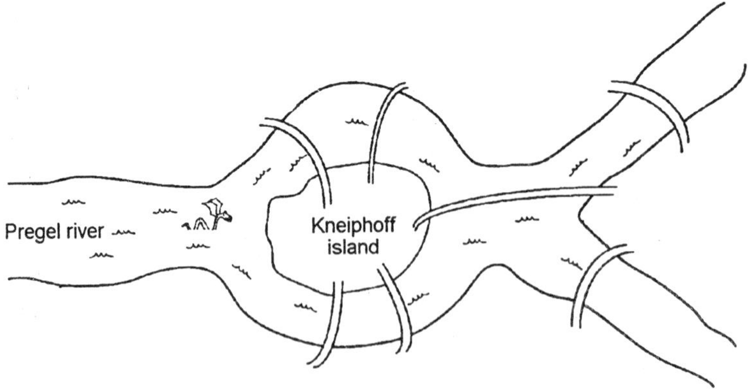

Fig. 8.7. Bridges of Konigsberg.

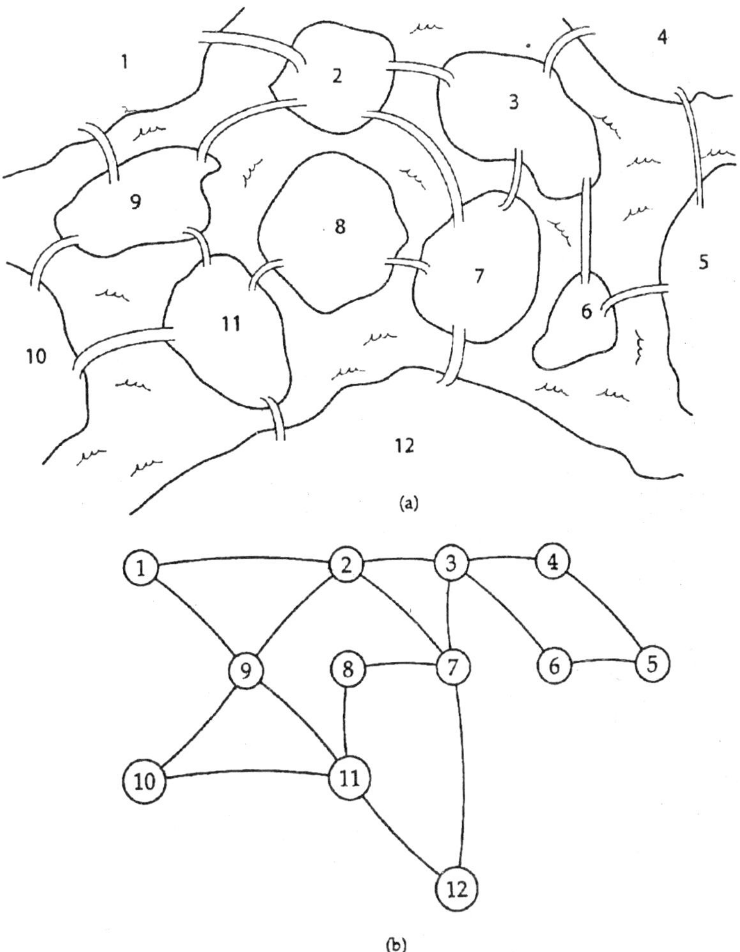

Fig. 8.8. A more complicated version of Konigsberg (a). To solve the Bridge obsession problem, Euler transformed the map of Knoigsberg into a graph (b) and found an Eulerian cycle.

transformation, the Bridge Obsession problem turns into the Eulerian Cycle problem that was solved by Euler and later found thousands of a applications in different areas of science and engineering: After the Konigsberg Bridge problem was solved, graph theory was forgotten for a century before it was rediscovered by Arthur Cayley who studied the chemical structures of (noncyclic) saturated hydrocarbons $C_n H_{2n+2}$. Structures of this type of hydrocarbon are examples of *trees*, which

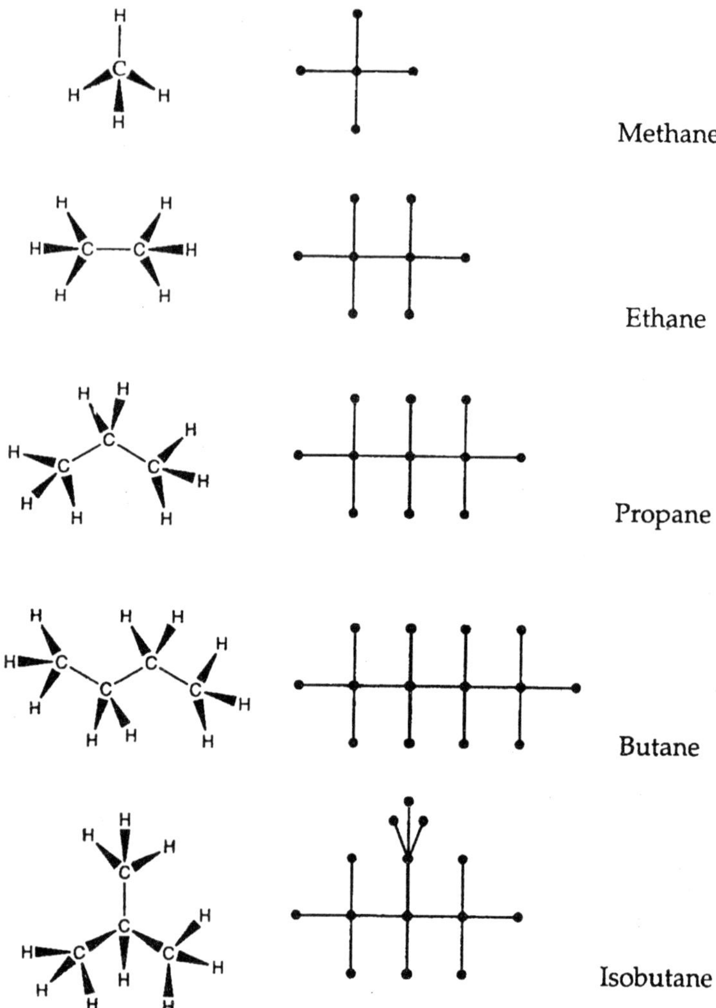

Fig. 8.9. Hydrocarbons as chemists see them (left) and their graph representation (right). Two different molecules with the same number of the same types of atoms are called structural isomers.

are simply connected graphs with no cycles. It is not hard to show that every tree has at least one vertex with degree 1, or *leaf*. This observation immediately implies that every tree on n vertices has $n - 1$ edges, regardless of the structure of the tree.

Indeed, since every tree has a leaf, we can remove it and its attached edge, resulting in another tree. So far we have removed

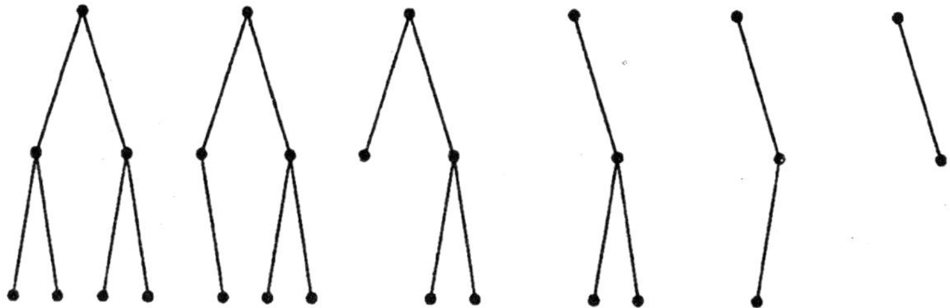

Fig. 8.10. Proving that a tree with n vertices has n – 1 edges.

one edge and one vertex. In this smaller tree there exists a leaf that we, again remove. So far, we have removed two vertices and two edges. We keep this up until we are left with a graph with a single vertex and no edges. Since we have removed $n - 1$ vertices and $n - 1$ edges, the number of edges in every tree is $n - 1$. Shortly after Cayley's work on tree enumeration, Sir William Hamilton invented a game corresponding to a graph whose twenty vertices were labeled with the names of twenty famous cities. The goal is to visit all twenty cities in such a way that every city is visited exactly once before returning back to the city where the tour started.

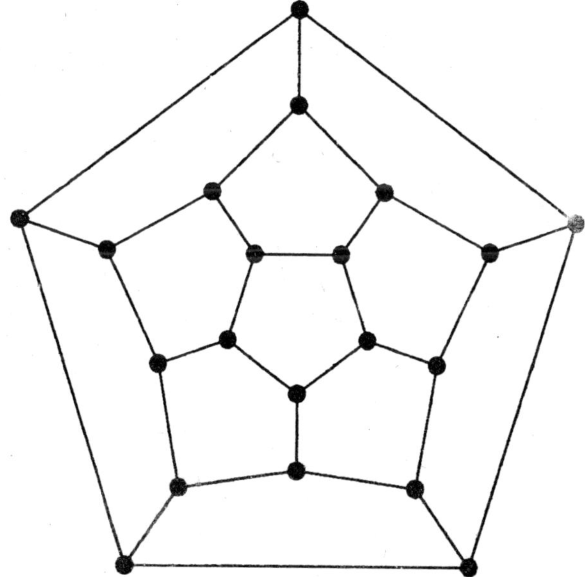

Fig. 8.11. Can you travel from any one of the vertices in this graph, visit every other vertex exactly once, and end up at the original vertex.

As the story goes, Hamilton sold the game for 25 pounds to a London game dealer and it failed miserably. Despite the commercial failure of a great idea, the more general problem of finding "Hamiltonian cycles" in arbitrary graphs is of critical importance to many scientific and engineering disciplines. The problem of finding Hamiltonian cycles looks deceivingly simple and somehow similar to the Eulerian Cycle problem. However, it turns out to be NP-complete while the Eulerian Cycle problem can be solved in linear time. Graphs, like many freeway maps, often give some sort of weight to every edge, as in the Manhattan Tourist problem.

The weight of an edge may reflect, depending on the context, different attributes. For example, the length of a freeway segment connecting two cities, the number of tourist attractions along a city block, and the alignment score between two amino acids are all natural weighting schemes. Weighted graphs are often formally represented as an ordered triple, $G = (V, E, w)$, where V is the set of vertices in the graph, E is the set of edges, and w is a weight function defined for every edge e in E(i.e., $w(e)$ is a number reflecting the weights of edge e). Given a weighted graph, one may be interested in finding some shortest path between two vertices. Though this problem may sound difficult if you were given a complicated road map, it turns out that there exist fast algorithms to answer this question.

We emphasize that this problem is different—and somewhat more complicated—than the Longest Path in a Directed Acyclic Graph (DAG). Graphs are powerful tools in many applied studies and their role goes well beyond analysis of maps of freeways. For example, at first glance, the following "basketball" problem has nothing to do with graph theory: "Fifteen teams play a basketball tournament in which each team plays with each other team. Prove that the teams can always be numbered from 1 to 15 in such a way that team 1 defeated team 2, team 2 defeated team 3, .., team 14 defeated team 15."

A careful analysis reduces this problem to finding a Hamiltonian path in a directed graph on fifteen vertices.

Graph and Genetics

Conceived by Euler, Cayley, and Hamilton, graph theory flourished in the twentieth century to become a critical component of discrete mathematics. In the 1950s, Seymour

Benzer applied graph theory to show that genes are linear. At that time, it was known that genes behaved as functional units of DNA, much like pearls on a necklace, but the chemical structures and organization of the genes was not clear. Were genes broken into still smaller components? If so, how were they organized?

Prior to Watson and Crick's elucidation of the DNA double helix, it seemed a reasonable hypothesis that the DNA content of genes was branched, or even looped, rather than linear. These two organizations have very different topological implications, which Benzer exploited in an ingenious experiment. Benzer studied a large number of bacteriophage T4 mutants, which happened to have a continuous interval deleted from an important gene. In their "normal" state, nonmutant T4 phages

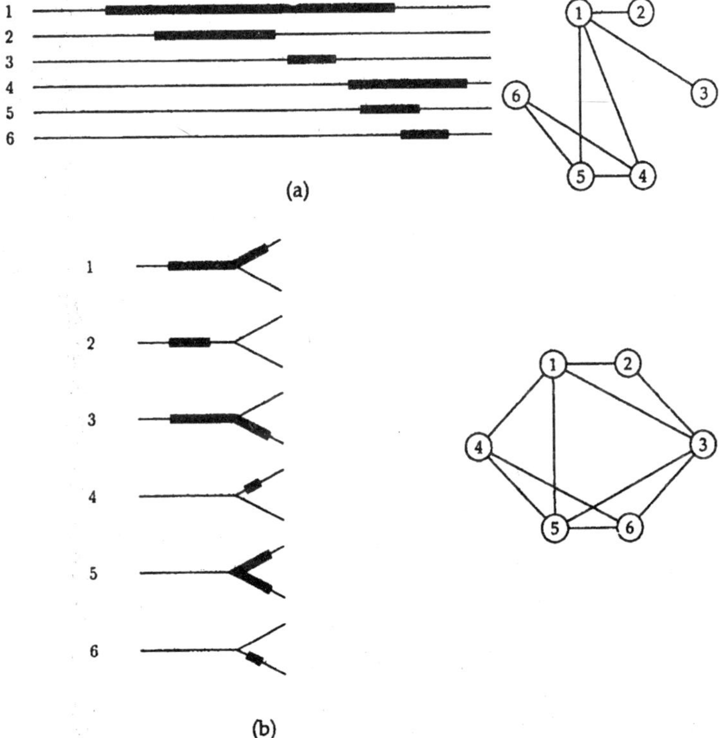

Fig. 8.12. Two different hypothetical structures of a gene. (a) Linear structures exhibit very different interval graphs than the graph exhibited by (b) branched structures.

will kill a bacterium. The mutant T4 phages that were missing a segment of their genome could not kill the bacterium. Different mutants had different intervals deleted from the gene, and Benzer had to determine which interval had been deleted in each mutant.

Though Benzer did not know exactly where in the gene the mutant's deletion was, he had away to test whether two deletions (i.e., their intervals) overlapped, relying on how two phages with two different deletions behave inside a bacterial host cell. When two mutant phages with two different deletions infect a bacterium, two outcomes are possible depending on whether the deleted intervals overlap or not. If they do not overlap, then two phages combined have all the genetic material of one normal phage. However, if two deleted intervals overlap, some genetic material is absent in both phages.

The Benzer experiment was based on the observation that in the former case (all genetic material of a normal phage is present), two mutants are able to ill the bacterium; and in the later case (where some genetic material is removed in both phages) the bacterium survives. Benzer infected bacteria with each *pair* of mutant strains from his T4 phage and simply noted which pairs killed the bacterial host. Pairs which were lethal to the host were mutants whose deleted intervals did not overlap. Benzer constructed a graph where each strain was a vertex and two vertices were connected when a double infection of a bacterial host was nonlethal. He reasoned that if genes were linear, he would probably see one type of graph, but if the genes were branched he would see another.

Given a set of n intervals on a line segment, the *interval graph* is defined as a graph with n vertices that correspond to the intervals. There is an edge between vertices v and w if and only if the intervals v and w overlap. Interval graphs have several important properties that make them easy to recognize. Benzer' problem was equivalent to deciding whether the graph obtained from his bacteriophage experiment represented an interval graph. Had the experiment resulted in a graph like the "house" graph, then the genes could not have been organized as linear structures. As it turned out, the graph was indeed an interval graph, indicating that genes were composed of linearly organized functional units.

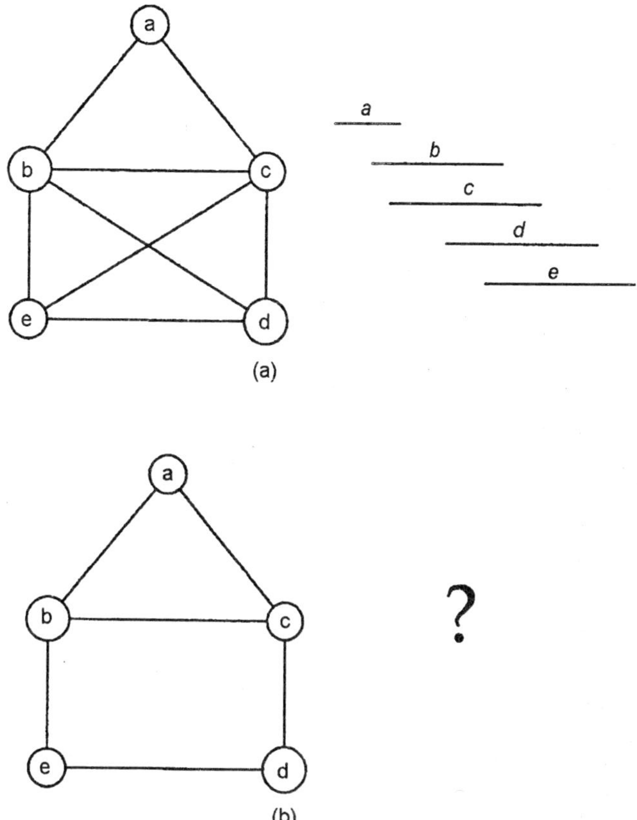

Fig. 8.13. An interval graph (a) and a graph that is not an interval graph (b).

DNA Sequencing

Imagine several copies of a magazine cut into millions of pieces. Each copy is cut in a different way, so a piece from one copy may overlap pieces from another. Assuming that some large number of pieces are just sort of lost, and the remaining pieces are splashed with ink, can you recover the original text? This essentially, is the problem of fragment assembly in DNA sequencing. Classic DNA sequencing technology allows one to read short 500- to 700-nucleotide sequences per experiment, each fragment corresponding to one of the many magazine pieces. Assembling the entire genome from these short fragments is like reassembling the magazine from the millions of tiny slips of paper.

Both problems are complicated by unavoidable experimental errors—ink splashes on the magazine, and mistakes in reading nucleotides. Furthermore, the data are frequently incomplete–some magazine pieces get lost, while some DNA fragments never make it into the sequencing machine. Nevertheless, efforts to determine the DNA sequence of organisms have been remarkably successful, even in the face of these difficulties. Two DNA sequencing methods were invented independently and simultaneously in Cambridge, England by Fred Sanger and in Cambridge, Massachusetts by Walter Gilbert. Sanger's method takes advantage of how cells make copies of DNA. Cells copy a strand of DNA nucleotide by nucleotide in a reaction that adds one base at a time.

Sanger realized that he could make copies of DNA fragments of different lengths if he starved the reaction of one of the four bases: a cell can only copy its DNA while it has all of the bases in supply. For a sequence ACGTAAGCTA, starving at T would produce a mixture of the fragments ACG and ACGTAAGC. By running one starvation experiment for each of A, T, G, and C and then separating the resulting DNA fragments by length, one can read the DNA sequences. Each of four starvation experiments produces a *ladder* of fragments of varying lengths called the *Sanger ladder*. This approach culminated in the sequencing of a 5386-nucleotide virus in 1977 and a Nobel Prize shortly thereafter.

Since then the amount of DNA sequence data has been increasing exponentially, particularly after the launch of the Human Genome Project in 1989. By 2001, it had produced the roughly 3 billion-nucleotide sequence of the human genome. Within the past twenty years, DNA sequencing technology has been developed to the point where modern sequencing machines can sequence 500- to 700-nucleotide DNA fragments, called *sequencing reads*. These reads then have to be assembled into a continuous genome, which turns out to be a very hard problem.

Even though the DNA reading process has become quite automated, these machines are not microscope-like devices that simply scan 500 nucleotides as if they were a sentence in a book. The DNA sequencing machine measure the lengths of DNA fragments in the Sanger ladder, but even this task is difficult; we cannot measure a *single* DNA fragment, but must measure billions of identical fragments. *Shotgun sequencing* starts with a large sample of genomic DNA. The sample is *sonicated*, a process which randomly partitions each piece of DNA in the sample into *inserts*; the inserts that are smaller than 500 nucleotides are removed from further consideration. Before the inserts can be read, each one must be multiplied billions of times so that it is possible to read the ladders produced by Sanger's technique.

To amplify the inserts, a sample is cloned into a *vector*, and this vector used to infect a bacterial host. As the bacterium reproduces, it creates a colony that contains billions of copies of the vector and its associated insert. As a result, the cloning process results in the production of a large sample of one particular insert that can then be sequenced by the Sanger method. Usually, only the first 500 to 700 nucleotides of the insert can be interpreted from this experiment. DNA sequencing is therefore a two stage process including both experimental (reading 500-700 nucleotide sequences from different inserts) and computational (assembling these reads into a single long sequence) components.

Shortest Superstring Problem

Since every string, or *read*, that we sequence came from the much longer genomic string, we are interested in a *superstring* of the reads—that is, we want a long string that "explains" all the reads we generated. However, there are many possible superstrings to choose from—for example, we could concatenate all the reads together to get a (not very helpful) superstring. We choose to be most interested in the shortest one, which turns out to be a reasonable first approximation to the unknown genomic DNA sequence. With this in mind, the simplest approximation of DNA sequencing corresponds to the following problems. Define $overlap(s_i, s_j)$ to be the length of the longest prefix of s_j that matches a suffix of s_i.

The shortest superstring problem can be cast as a Traveling Salesman problem in a complete directed graph with n vertices corresponding to strings $s_1, ..., s_n$ and edges of length $-overlap(s_i, s_j)$. This reduction, of course, does not lead to an efficient algorithm since the TSP is *NP*-complete. Moreover it is known that shortest superstring problem is itself *NP*-complete, so that a polynomial algorithm for this problem is unlikely. The early DNA sequencing algorithms used a simple greedy strategy: repeatedly merge a pair of strings with maximum overlap until

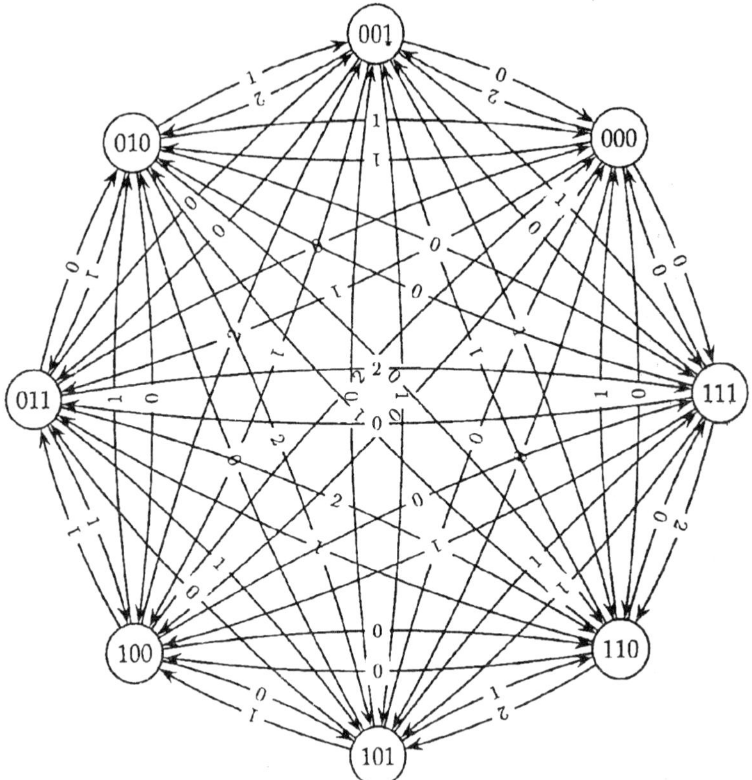

Fig. 8.14. The overlap graph for the eight strings.

only one string remains. It has been conjectured but not yet proved, that this greedy algorithms has performance guarantee 2.

DNA Arrays as an Alternative Sequencing Technique

When the Human Genome Project started, DNA sequencing was a routine but time-consuming and hard-to-automate procedure. In 1988 four groups of biologists independently and simultaneously suggested a different sequencing technique called *Sequencing by Hybridization*. SBH involves building a miniature *DNA array*, also known as *DNA chip*, that contains thousands of short DNA fragments called *probes*. Each of these short fragments reveals. Whether or not a known—but short—sequence occurs in the unknown DNA sequence; all these pieces of information together should reveal the identify of the target DNA sequence.

Given a short probe (an 8- to 30-nucleotide single-stranded synthetic DNA fragments) and a single-stranded target DNA fragment, the target will *hybridize* with the probe if the probe is a substring of the target's Watson-Crick *complement*.

When the probe and the target are mixed together, they form a weak chemical bond and stick together. For example, a probe ACCGTGGA will hybridize to a target CCTGGCACCTA since its is complementary to the substring TGGCACCT of the target. In 1988 almost nobody believed that the idea of using DNA probes to sequence long genomes would work, because both the biochemical problem of synthesizing thousands of short DNA fragments and the combinatorial problem of sequence reconstruction appeared too complicated. Shortly after the first paper describing DNA arrays was published, the journal *Science* wrote that given the amount of work required to synthesize a DNA array, "[using DNA arrays for sequencing] would simply be substituting one horrendous task for another."

A major breakthrough in DNA array technology was made by Steve Fodor and colleagues in 1991. Their approach to array manufacturing relies on *light-directed polymer synthesis*, which has many similarities to computer chip manufacturing. Using this technique, building an array with all 4^l probes of length l requires just $4 \cdot l$ separate reactions, rather than the presumed 4^l reactions. With this method, a California-based biotechnology company, Affymetrix, built the first 64-kb DNA array in 1994.

Today, building 1-MB or larger arrays is routine, and the use of DNA arrays has become one of the most widespread new biotechnologies. SBH relies on the hybridization of the target DNA fragment against a very large array of short probes. In this manner, probes can be used to test the unknown target DNA to determine its *l-mer composition*. The *universal* DNA array contains all 4^l probes of length l and is applied as follows:

1. Attach all possible probes of length l ($l = 8$ in the first SBH papers) to a flat surface, each probe at a distinct and known location. This set of probes is called the *DNA array*.
2. Apply a solution containing fluorescently labeled DNA fragment to the array.
3. The DNA fragment hybridizes with those probes that are complementary to substrings of length l of the fragment.

Universal DNA Array

	AA	AT	AG	AC	TA	TT	TG	TC	GA	GT	GG	GC	CA	CT	CG	CC
AA																
AT			ATAG													
AG																
AC												ACGC				
TA											TAGG					
TT																
TG																
TC																
GA																
GT																
GG													GGCA			
GC	GCAA															
CA	CAAA															
CT																
CG																
CC																

DNA target TATCCGTTT (complement of ATAGGCAAA) hybridizes to the array of all 4-mers:

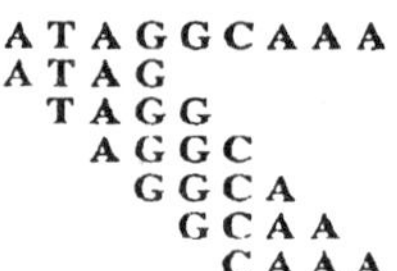

Fig. 8.15. Hybridization of TATCCGTTT with the universal DNA array consisting of all 4^4 4-mers.

4. Using a spectroscopic detector, determine which probes hybridize to the DNA fragment to obtain the l-mer composition of the target DNA fragment.
5. Apply the combinatorial algorithm described below to reconstruct the sequence of the target DNA fragment from the l-mer composition.

Sequencing by Hybridization

Given an unknown DNA sequencing, an array provides information about all strings of length l that the sequence contains, but does not provide information about their positions in the sequence. For a string s of length n, the l-mer composition, or *spectrum*, of s, is the multiset of $n - l + 1$ l-mers in s and is written *Spectrum* (s, l). If $l = 3$ and s = TATGGTGC, then *Spectrum* (s, l) = {TAT, ATG, TGG, GGTS, GTG, TGC}. We can now formulate the problem of sequencing a target DNA fragment from its DNA array data. Although conventional DNA sequencing and SBH are very different experimental approaches, you can see that the corresponding computational problems are quite similar.

In fact, SBH is a particular case of the Shortest Superstring problem when the strings s_1, ..., s_n represent the set of all sub strings of s of fixed size. However, in contrast to the Shortest Superstring problem, there exists a simple linear-time algorithm for the SBH problem. Notice that it is not a contradiction that the shortest superstring problem is *NP*-complete, yet we claim to have a linear-time algorithm for the SBH problem, since the Shortest Superstring problem is more general than the SBH problem. Although DNA arrays were originally proposed as an alternative to conventional DNA sequencing, do novo sequencing with DNA arrays remains an unsolved problem in practice.

The primary obstacle to applying DNA arrays for sequencing is the inaccuracy in interpreting hybridization data to distinguish between perfect matches [i.e. l-mers present in *Spectrum*(s, l)] and highly stable mismatches (i.e. l-mers not present in *Spectrum*(s, l), but with sufficient chemical bonding potential to generate a strong hybridization signal). This is a particularly difficult problem for the short probes used in universal arrays. As a result, DNA arrays have become more popular in gene expression analysis and studies of genetic variations—both of which are done with longer probes—than in *de novo* sequencing.

In contrast to SBH where the target DNA sequence is unknown, these approaches assume that the DNA sequence is either known or "almost" known (i.e, known up to a small number of mutations). For example to detect genetic variations one can design twenty-to thirty-nucleotide probes to reliably detect mutations, bypassing the still unsolved problem of distinguishing perfect matches from highly stable mismatches in the case of short probes. To detect mutations in the (known) sequence *S*, an array should contain all 25-mers from *S*, as well a selected mutated versions of these 25-mers.

SBH as a Hamiltonian Path Problem

Two l-mers p and q *overlap* if $overlap(p, q) = l - 1$, that is, the last $l-1$ letters of p coincide with the first $l-1$ letters of q. Given the measured spectrum *Spectrum*(s, l) of a DNA fragment s, construct a directed graph, H, by introducing a vertex for every l-mer in *Spectrum*(s, l), and connect every two vertices p and q by the directed edge (p, q) if p and q overlap. There is a one-to-one correspondence between paths that visit each vertex of H exactly once and DNA fragments with the spectrum *Spectrum*(s, l). The spectrum corresponds to the sequence reconstruction ATGCAGGTCC, which is the only path visiting all vertices of H:

ATG → TGC → GCA → CAG → AGG → GGT → GTC → TCC

The spectrum yields a more complicated graph with two Hamiltonian path, each path corresponding to two possible reconstruction: ATGCGTGGCA and ATGGCGTGCA. As the overlap graph becomes larger, this approach ceases to be practically useful since the Hamiltonian Path problem is *NP*-complete.

SBH as an Eulerian Path Problem

As we have seen, reducing the SBH problem to a Hamiltonian Path problem does not lead to an efficient algorithm. Fortunately, reducing SBH to the *Eulerian Path* problem in a directed graph, which leads to the simple linear-time algorithm for sequence reconstruction mentioned earlier. The reduction of the SBH problem to an Eulerian Path problem is to construct a graph whose edges—rather than vertices—correspond to l-mers from *Spectrum*(s, l) and then to find a path in this graph visiting every edge exactly once. In this approach we build a graph G on the set of all $(l - 1)$-mers, rather than on the set of all l-mers as in the previous section. An $(l - 1)$ mer v is joined by a directed edge with an $(l - 1)$-mer w if the spectrum contains an l-mer for which the first $l - 1$ nucleotides coincide with v and the last $l - 1$ nucleotides coincide with w. Each l-mer from the spectrum corresponds to a directed edge in G rather than to a vertex as it does in H. Therefore, finding a DNA fragment containing all l-mers from the spectrum corresponds to finding a path visiting all *edges* of G, which is the problem of finding an Eulerian path.

Superficially finding an Eulerian path looks just as hard as finding a Hamiltonian path, but, as we show below, finding Eulerian paths turns out to be simple. We will first consider

Eulerian cycles, that is, Eulerian paths in which the first and the last vertices are the same. A directed graph G is *Eulerian* if it contains an Eulerian cycle. A vertex v in a graph is *balanced* if the number of edges entering v equals the number of edges leaving v, that is, if $indegree(v) = outdegree(v)$. For any given vertex v is an Eulerian graph, the number of times the Eulerian cycle enters v is exactly the same as the number of times it leaves v. Thus, is, if $indegree(v) = outdegree(v)$ for every vertex v is an Eulerian graph, motivating the following theorem characterizing Eulerian graphs.

Theorem. A connected graph is Eulerian if the an only if each of it vertices is balanced.

Proof. First, it si easy to see that if a graph is Eulerian, then each vertex must be balanced, We show that if each vertex is a connected graph is balanced, then the graph is Eulerian.

To construct an Eulerian cycle, we start from an arbitrary vertex v and form any arbitrary path by traversing edges that have not already been used. We stop the path when we encounter a vertex with no way out, that is, a vertex whose outgoing edges have already been used in the path. In a balanced graph, the only vertex where this can happen is the starting vertex v since for any other vertex, the balance condition ensure that for every incoming edge there is an outgoing edge that has not yet been used. Therefore, the resulting path will end at the same vertex where it started, and with some luck will be Eulerian. However, if the path is not Eulerian, it must contain a vertex w that still has some number of untraversed edges.

The cycle we just constructed forms a balanced subgraph. Since the original graph was balanced, then the edges that were not traversed in the first cycle also form a balanced sub-graph. Since all vertices in the graph with untraversed edges are balanced there must exist some other path starting and ending at w, containing only untraversed edges. One can now combine the two paths into a single one as follows. Traverse the first path from v to w, then traverse the second path from w back to itself, and then traverse the remainder of the first path w back to v. Repeating this until there are no more vertices with unused edges will eventually yield an Eulerian cycle. This algorithm can be implemented in time linear in the number of edges in the graph.

Notice that we have described Eulerian graphs as containing an Eulerian *cycle*, rather than an Eulerian path, but we have said

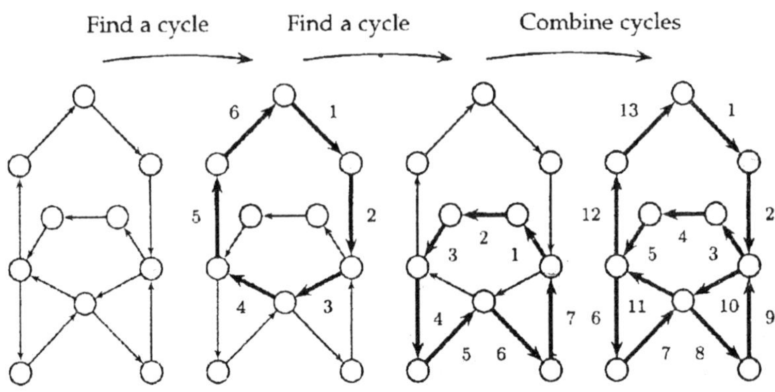

Fig. 8.16. Constructing an Eulerian cycle.

that the SBH problem reduces to that of finding an Eulerian path. A vertex v in a graph is called *semibalanced* if $|indegree(v) - outdegree(v)| = 1$. If a graph has an Eulerian path starting at vertex s and ending at vertex t, then all its vertices are balanced, with the possible exception of s and t, which may be semibalanced.

The Eulerian path problem can be reduced to the Eulerian cycle problem by adding an edge between two semibalanced vertices. This transformation balances all vertices in the graph and therefore guarantees the existence of an Eulerian cycle in the graph with the added edge. Removing the added edge form the Eulerian cycle transforms it into an Eulerian path. The following theorem characterizes all graphs that contain Eulerian paths.

Theorem. A connected graph has an Eulerian path if and only if it contains at most two semibalanced vertices and all other vertices are balanced.

Fragment Assembly in DNA Sequencing

As we mentioned previously, after the short 500- to 700-bp *DNA reads* are sequenced, biologists need to assemble them together to reconstruct the entire genomic DNA sequence. This is known as fragment assembly. The shortest superstring problem described above is an overly simplified abstraction that does not adequately capture the essence of the fragment assembly problem, since it assumes error-free read. The error rate in DNA reads produced by modern sequencing machines varies from 1% to 3%. A further complication in fragment assembly in the fact that one does not know a priori which of two DNA strands a read came from. DNA is double-stranded, and which of the two strands was sequenced by a read depends on how the insert was oriented in the vector. Since this is essentially arbitrary, one never knows whether a read came from a target strand DNA sequence or from its Watson-Crick complement. However, sequencing errors and assignments of reads to one of two strands are just minor annoyances compared to the major problem in fragment assembly: repeats in DNA.

The human genome contains many sequences that repeat themselves throughout the genome a surprisingly large number of times. For example, the roughly 300 nucleotide *Alu* sequence is repeated more than a million times throughout the genome, with only 5% to 15% sequence variation. Even more troublesome for fragment assembly algorithms is the fact that repeats occur at several scales. The human T-cell receptor locus contains five closely located repeats of the trypsinogen gene, which is 4 kb long and varies only by 3% to 5% between copies. These long repeats are particularly difficult to assemble, since there are no reads with unique portions flanking the repeat region. The human genome contains more than 1 million *Alu* repeats (≈ 300 bp) and 200,000 *LINE* repeats (≈ 100 bp), not to mention that an estimated 25% of genes in the human genome have duplicated copies.

A little arithmetic shows that these repeats and duplicated genes represent about half the human genome. If one models the genome as a 3 billion-letter sequence produced by a random number generator, then assembling it from 500-letter reads is actually relatively simple. However because of the large number of repeats, it is theoretically impossible to uniquely assemble real reads as long as some repeats are longer than the typical read length. Increasing the length of the reads (to make them longer than most repeats) would solve the problem, but the sequencing technology has not significantly improved the read length yet. A puzzle that looks deceivingly simple and has only sixteen triangular pieces. People usually assemble puzzle by connecting matching pieces. In this case, for every triangle in the puzzle, there is variety of potentially matching triangles (every frog in the puzzle is repeated several times). As a result, you cannot know which of the potentially matching triangles is the correct one to use at any step. If you proceed without some sort of guidance, you are likely to end up in the situation. Fourteen of the pieces have been placed completely consistently, but the two remaining pieces are impossible to place.

It is difficult to design a strategy that can avoid such dead ends for this particular puzzle, and it is even more difficult to design strategies for the linear puzzle presented by repeats in a genome. Since repeats present such a challenge in assembling long genomes, the original strategy for sequencing the human genome was first to clone it into BACs, each BAC carrying an approximately 150,000 bp long insert. After constructing a library of overlapping BACs that covers the entire human genome (which requires approximately 30,000 BACs), each one can be sequenced as if it were a separate minigenome. This BAC-by-BAC sequencing strategy significantly simplifies the computational assembly problem (by virtue of the fact that the number of repeats present within a BAC is 30,000 times smaller than the number of repeats in the entire genome) but makes the sequencing project substantially more cumbersome.

Although the Human Genome project demonstrated that this BAC-by-BAC strategy can be successful, recent large-scale sequencing projects (including the 2002 mouse genome assembly) mainly follow the whole-genome assembly paradigm advocated by James Weber and Gene Myers in 1997. Myers led the fragment assembly efforts at Celera Genomics, the company that announced the completion of a (draft) human genomic sequence in 2001. Weber and Myers suggested a virtual increase in the length of a read, by pairing reads that were separated by a fixed-size gap. This suggesting resulted in the so-called *mate-pair reads* sequencing technique.

In this method, inserts of length approximately L (where L is much longer than the length of a read) are selected, an *both* ends of the insert are sequenced. This produces a pair of reads called mates at a known (approximate) distance, L, from each other. The insert length L is chosen in a such away that it is larger than the length of most repeats in the human genome. The advantage of mate-pair reads is that it is unlikely that both reads of the mate-pair will lie in a large-scale DNA repeat. Thus the read that lies in a unique portion of DNA determines which copy of a repeat its mate is in. Most fragment assembly algorithms consist of the following three steps:

- *Overlap*: Finding potentially overlapping reads.
- *Layout*: Finding the order of reads along DNA.
- *Consensus*: Deriving the DNA sequence form the layout.

The overlap problem is to find the best match between the suffix of one read and the prefix of another. In the absence of sequencing errors, we could simply find the longest suffix of one string that exactly matches the prefix of another string. However, sequencing errors force us to use a variation of the dynamic programming algorithm for sequence alignment. Since errors are small (1% to 3%), the common practice is to filter out pairs of fragments that do not share a significantly long common substring. Constructing the layout is the hardest step in fragment assembly. The difficulty is in deciding whether two fragments really overlap (i.e., their differences are caused by sequencing errors) or actually come from two different copies of repeat.

Repeats represent a major challenge for whole-genome shotgun sequencing and make the layout problem very difficult. The final consensus step of fragment assembly amounts to correcting errors in sequence reads. The simplest way to build the consensus is to report the most frequent character in the layout constructed in the layout step. This assumes that each position in the genome was represented by a sufficiently large number of reads to ensure that experimental errors are reduced to minor noise.

PROTEIN SEQUENCING AND IDENTIFICATION

Few people remember that before DNA sequencing had even been seriously suggested, scientists routinely sequenced proteins. Frederick Sanger was awarded his first (of two) Nobel prize for determining the amino acid sequence of insulin, the protein needed by people suffering from diabetes. Sequencing the 52-amino acid bovine insulin in the late 1940s seemed more challenging than sequencing an entire genome seems today. The computational problem facing protein sequencing at that time was similar to that facing modern DNA sequencing; the main difference was in the length of the sequenced fragments.

In the late 1940s, biologists discovered how to apply the *Edman degradation reaction* to chop off one terminal amino acid at a time from the end of a protein and read it. Unfortunately, this only works for a few terminal amino acids before the results become impossible to interpret. To get around this problem, Sanger digested insulin with *proteases* (enzymes that cleave proteins) into *peptides* (short protein fragments) and sequenced each of the resulting fragments independently. He then used these overlapping fragments to reconstruct the entire sequence, exactly like the DNA sequencing "break—read the fragments—assemble" method today.

The Edman degradation reaction became the predominant protein sequencing method for the next twenty years, and by the late 1960s, protein sequencing machines were on the market. Despite these advances, protein sequencing ceased to be of central interest in the field as DNA sequencing technology underwent rapid improvement in the late 1970s. In DNA sequencing,

GIVE
GIVEECCA
GIVEECCASV
GIVEECCASVC
GIVEECCASVCSL
GIVEECCASVCSLY
SVC
SLY
SLYELEDYC
YE
YEL
YELE
ELEDY
ELEDYCD
LE
LEDYCD
EDYCD
DYCD
CD

FVDEHLCG
FVDEHLCGSHL
HLCGSHL
SHLVEA
VEAL
VEALY
AL
ALY
YLVCG
LVCGERGF
LVCGERGFF
GERG
GF
GFFYTPK
YTPKA
TPKA

Fig. 8.17. The peptide fragments that Frederick Sanger obtained from insulin through a variety of methods.

obtaining reads is relatively easy; it is the assembly that is difficult. In protein sequencing, obtaining reads is the primary problem, while assembly is easy. Having DNA sequence data for a cell is critical to understanding the molecular processes that the cell goes through. However, it is not the only important component: one also needs to know what proteins the cell produces and what they do. On the one hand, we do not yet know the full set of proteins that cells produce, so we need way to discover the sequence of previously unknown proteins. On the other hand, it is important to identify which specific proteins interact in a biological system (e.g., these proteins involved in DNA replication).

Lastly, different cells in an organism have different repertoires of expressed proteins. Brain cells need different proteins to function than liver cells do, and an important problem is to identify proteins that are present or absent in each biological tissue under different conditions. There are two types of computational problems motivated by protein sequencing. *De novo protein sequencing* is the elucidation of a protein's sequence in the case when a biological sample contains a protein that is either not present in a database or differs from a canonical version present in a database (e.g., mutated proteins or proteins with biochemical modifications).

The other problem is the identification of a protein that is present in a database; this is usually referred to as *protein identification*. The main difference between protein sequencing algorithms and protein identification algorithms is the difficulty of the underlying computational problems. Perhaps the easiest way to illustrate the distinction between protein identification and sequencing is with a gedanken experiment. Suppose a biologist wants to determine which proteins from the DNA polymerase complex in rats. Having the complete rate genome sequence and knowing the location of all the rat genes does not yet allow a biologist to determine what chemical reactions occur during the DNA replication process. However, isolating a rat's DNA polymerase complex, breaking it apart, and sequencing the proteins that form parts of the complex will yield a fairly direct answer to the researcher's question.

Of course, if we presume that the biologist has the complete rat genome sequence and all of its gene *products*, he may not actually have to sequence every amino acid in every protein in the DNA polymerase complex—just enough to figure out which proteins are present. This is protein identification. On the other hand, if the researcher decides to study an organism for which complete genome data are not available (perhaps an obscure species of ant), then the researcher will need to perform de novo protein sequencing. For many problems, protein sequencing and identification remains the only ways to probe a biological process. For example gene splicing is a complex process performed by the large molecular complex called the *spliceosome*, which consists of over 100 different proteins complexed with some functional RNA. Biologists want to determine the "parts list" of the spliceosome, that is, the identity of proteins that form the complex. DNA sequencing is not capable of solving this problem directly: even if all the proteins in the genome were known, it is not clear which of them are parts proteins in the genome were known, it is not clear which of them are parts of the spliceosome.

Protein sequencing and identification, on the other hand, are very helpful in discovering this parts list. Recently, Matthias Mann and colleagues purified the spliceosome complex and used protein sequencing and protein identification techniques to find a detailed parts list for it. Another application of these technologies is the study of proteins involved in *programmed cell death*, or *apoptosis*. In the development of many organisms cells must die at specific times. A cell dies if it fails to acquire certain survival factors, and the death process can be initiated by the expression of certain genes. In a developing nematode, for example, the death of individual cells in the nervous system may be prevented by mutations in several genes that are the subject of active investigation. DNA sequence data alone are not sufficient to find the genes involved in programmed cell death, and until recently, nobody knew the identity of these proteins.

Protein analysis by *mass spectrometry* allowed the sequencing of proteins involved in programmed cell death, and the discovery of some proteins involved in the death-including signaling complex. The exceptional sensitivity of mass spectrometry has opened up new experimental and computational possibilities of protein studies. A protein can be digested into peptides by proteases like trypsin. In a matter of seconds, a *tandem mass spectrometer* breaks a peptide into even smaller fragments and measures the mass of each. The *mass spectrum* of a peptide is a collection of masses of these fragments.

The protein sequencing problem is to derive the sequence of a peptide given its mass spectrum. For an ideal fragmentation process where every fragment of a peptide is generated, an in an ideal mass spectrometer, the peptide sequencing problem is simple. However, the fragmentation process is not ideal, and mass spectrometers measures mass with some imprecision. These details make peptide sequencing difficult. A mass spectrometer works like a charged sieve. A large molecule (peptide) gets broken into smaller fragments that have an electrical charge. These fragments are then spun around and accelerated in a magnetic field until they hit a detector. Because large fragments are harder to spin than small ones, one can distinguish between fragments with different masses based on the amount of energy required to fling the different fragments around. It happens that most molecules can be broken in several places, generating several different *ion types*. The problem is to reconstruct the amino acid sequence of the peptide from the masses of these broken pieces.

Peptide Sequencing Problem

Let $A = \{a_1, a_2, \ldots a_{20}\}$ be the set of amino acids, each with molecular masses $m(a_i)$. A *peptide* $P = p_1 \ldots p_n$ is a sequence of amino acids, with *parent* mass $m(P) = \Sigma^n_{i=1} m(p_i)$. We will denote the partial *N-terminal* peptide $p_1, \ldots p_i$ of mass $m_i = \Sigma^i_{j=1} m(p_j)$ as P_i and partial *C-terminal* peptide $p_{i+1}, \ldots, p_n$ of mass $m(P) - m_i$ as P_i^-, for $1 \leq i \leq n$. Mass spectra obtained by tandem mass spectrometry (MS/MS) consist predominantly or partial N-terminal peptides and C-terminal peptides.

A mass spectrometer typically breaks a peptide $p_1p_2 \ldots p_n$ at different peptide bonds and detects the masses of the resulting partial N-terminal and C-terminal peptides. For example, the peptide GPFNA may be broken into the N-terminal peptides G, GP, GPF, GPFN, and C-terminal peptides PFNA, FNA, NA, A. Moreover, while breaking GPFNA into GP and FNA, it may lose some small parts of GP and FNA, resulting in fragments of a lower mass. For example, the peptide GP might lose a water (H_2O), and the peptide FNA might lose an ammonia (NH_3). The resulting masses detected by the spectrometer will be equal to the mass of GP minus the mass of water (water happens to weight 1 + 1 + 16 = 18 daltons), and the mass of FNA minus the mass of ammonia

(1 + 1 + 1 + 14 =17 daltons). Peptides missing water and ammonia are two different *ion types* that can occur in fragmenting a peptide in a mass spectrometer.

Peptides fragmentation in a tandem mass spectrometer can be characterized by a set of numbers $\Delta = \{\delta_1, ..., \delta_k\}$ representing the different types of ions that correspond to the removal of a certain chemical group from a peptide fragment. We will call Δ the set of ion types. A *δ-ion* of N-terminal partial peptide P_i is a modification of P_i that has mass $m_i = \delta$, corresponding to the loss of a (typically small) chemical group of mass δ when *P* was fragmented into P_i. The *δ-ion* of C-terminal peptides is defined similarly. The most frequent N-terminal ions are called *b*-ions (ion b_i corresponds to P_i with $\delta = -1$) and the most frequent C-terminal ions are called *y*-ions (ion y_i corresponds to P_i^- with $\delta = 19$). Examples of other frequent N-terminal ions are represented by *b*-H_2O (a *b*-fragment that loses a water) or *y*-NH_3 and some others like *b*-H_2O-NH_3.

For tandem mass spectrometry, the *theoretical* spectrum *T*(*P*) of peptides *P* can be calculated by subtracting all possible ion types $\delta_1, ..., \delta_k$ from the masses of all partial peptides of *P*, such that every partial peptide generates *k* masses in the theoretical spectrum. An *experimental* spectrum $S = \{s_1, ..., s_q\}$ is a set of numbers obtained in a mass spectrometry experiment that includes masses of some fragment ions as well as chemical noise. Note that the distinction between the theoretical spectrum *T*(*P*) and the experimental spectrum *S* is that you mathematically generate *T*(*P*) given the peptide sequence *P*, but you experimentally generate *S* without knowing what the peptide sequence is that generated it. The *match* between the experimentally measured spectrum *S* and peptide *P* is the number of masses in *S* that are equal to masses in *T*(*P*). This is often referred to as the *shared peaks count*. In reality, peptide sequencing algorithms use more sophisticated objective functions than a simple shared peaks count, incorporating different weighting functions for the matching masses. We formulate the Peptide Sequencing problem as follows. In reality, mass spectrometers measure both mass and *intensity*, which reflects the number of fragment ions of a given mass are detected in the mass spectrometer. As a result, mass spectrometrists often represent spectra in two dimensions and refer to the masses in the spectrum as "peaks."

SPECTRUM GRAPHS

There are two main approaches to solving the Peptide Sequencing problem that researchers have tried: either through exhaustive search among all amino acid sequences of a certain length, or by analyzing the *spectrum graph* which we define below. The former approach involves the generation of all 20^l amino acid sequence of length l and their corresponding theoretical spectra, with the goal of finding a sequence with the best match between the experimental spectrum and the sequence's theoretical spectrum. Since the number of sequences grows exponentially with the length of the peptide, different branch-and-bound techniques have been designed to limit the combinatorial explosion in these methods.

Prefix pruning restricts the computational space to sequences whose prefixes match the experimental spectrum well. The difficulty with the prefix pruning is that it frequently discards the correct sequence if its prefixes are poorly represented in the spectrum. The spectrum graph approach, on the other hand, does not involve generating all amino acid sequences, and leads to a fast algorithm for peptide sequencing. In this approach, we construct a graph from the experimental spectrum. Assume for simplicity that an experimental spectrum $S = \{s_1, ..., s_q\}$ consists of N-terminal ions and we will ignore the C-terminal ions for a while. Every mass $s \in S$ may have been created form a partial peptide by one of the k different ion types. Since we do not know which ion type from $\Delta = (\delta_1, ..., \delta_k)$ created the mass s in the experimental spectrum, we generate k different "guesses" for each of masses in the experimental spectrum.

Every guess corresponds to the hypothesis that $s = x - \delta_j$, where x is the mass of some partial peptide and $1 \leq j \leq k$. Therefore, for every mass s in the experimental spectrum, there are k guesses for the mass x of some partial peptide: $s + \delta_1$, $s + \delta_2$, ..., $s + \delta_k$. As a result, each mass in the experimental spectrum is transformed into a set of k vertices in the spectrum graph, one for each possible ion type. The vertex for δ_i for the mass s is labeled with mass $s + \delta_i$. We connect any two vertices u and v in the graph by the directed edge (u, v) if the mass of v is larger than that of u by the mass of a single amino acid. If we add a vertex at 0 and a vertex at the parent mass m (connecting them to other vertices as before), then the Peptide Sequencing problem can be cast as finding a path from 0 to m in the resulting DAG. In summary, the vertex set of the resulting spectrum graph is a set of numbers $s_i + \delta_j$ representing potential masses of N-terminal peptides adjusted by the ion type δ_j. Every mass s_i of spectrum S generates k distinct vertices $V_i(s) = \{s_i + \delta_1, ..., s_i + \delta_k\}$, though the sets V_i and V_j may overlap if s_i and s_j are close. The set of vertices in a spectrum graph is therefore $\{s_{initial}\} \cup V_1 \cup ... \cup V_q \cup \{s_{final}\}$, where $s_{initial} = 0$ and $s_{final} =$ m.

The spectrum graph may have a most $q_k + 2$ vertices. We label the edges of the spectrum graph by the amino acid whose mass is equal to difference between vertex masses. If we look at vertices as putative N-terminal peptides, the edge form u to v implies that the N-terminal sequence corresponding to v may be obtained by extending the sequence at u by the amino acid that labels (u, v). A spectrum S of a peptide $P = p_1...p_n$ is called *complete* if S contains a least one ion type corresponding to every N-terminal partial peptide P_i for every $1 \leq i \leq n$. The use of a spectrum graph is based on the observation that for a complete spectrum there exists a path of length $n + 1$ from $s_{initial}$ to s_{final} in the spectrum graph that is labeled by P.

This observation casts the Peptide Sequencing problem as one of finding the "correct" path in the set of all paths between two vertices in a directed acyclic graph. If the spectrum is complete, then the correct path that we looking for is often the path with the maximum number of edges, the familiar Longest Path in a DAG problem.. Unfortunately, experimental spectra are frequently incomplete. Moreover, even if the experimental spectrum is complete, there are often many paths in the spectrum graph to choose from that have same (or even larger) length, preventing one from unambiguously reconstructing the peptide.

The problem with choosing a path with a maximum number of edges is that it dose not adequately reflect the "importance" of different vertices. For example, a vertex in the spectrum graph obtained by a shift of +1 as $s_i + 1$ (corresponding to the most frequent b-ions) should be scored higher than a vertex obtained by a shift of the rate b_2-H_2O-NH_3 ion ($s_i + 1 - 18 - 17 = s_i - 34$). Further, whenever there are two peaks s_i and s_i' such that $s_i + \delta_j = s_i' + \delta_{j'}$, the vertex corresponding to that mass should also get a higher score than a vertex obtained by a single shift. In the probabilistic approach to peptide sequencing, each ion type δ_i has some probability of occurring which we write as $p(\delta_i)$. Under the simplest assumption, the probability that δ_i occurs for some partial peptide is independent of whether δ_j also occurs for the same partial peptide.

Under this assumption, any given partial peptide may contribute as many as k masses in the spectrum [this happens with probability $\Pi^k_{i=1} p(\delta_i)$] and as few as 0 [this happens with probability $\Pi^k_{i=1} (1 - p(\delta_i))$]. The probabilistic model below scores the vertices of the spectrum graph based on these simple assumptions. Suppose that an N-terminal partial peptide P_i with mass m_i produces ions $\delta_i, ..., \delta_i$ ("present" ions of mass $m_i - \delta_i, ..., m_i - \delta_l$) but fails to produce ions $\delta_i + 1, ..., \delta_k$ ("missing" ions) in the experimental spectrum. All l present ions will result in a vertex in the spectrum graph at mass m_i, corresponding to P_i. How should we score this vertex? A naive approach would be to reward P_i for every ions type that explains it, suggesting a score of $\Pi^l_{i=1} p(\delta_i)$. However, this approach has the disadvantage of not considering the missing ions, so we combine those in by defining the score of the partial peptide to be

$$\left(\prod_{i=1}^{l} p(\delta_i)\right)\left(\prod_{i=l+1}^{k} (1 - p(\delta_i))\right)$$

However, there is some inherent probability of chemical noise, that is, it can produce *any* mass (that has nothing to do with a peptide of interest) with certain probability p_R. Therefore, we adjust the probabilistic score as

$$\left(\prod_{i=1}^{i} \frac{p(\delta_i)}{p_R}\right)\left(\coprod_{i=l+1}^{k} \frac{1 - p(\delta_i)}{1 - p_R}\right)$$

Protein Identification via Database Search

De novo protein sequencing algorithms are invaluable for identification of (both known and unknown) proteins, but they are most useful when working with complete or nearly complete high-quality spectra. Many spectra are far from complete, and de novo peptide sequencing algorithms often produce ambiguous solutions for such spectra. If we had access to a database of all proteins from a genome, then we would not longer need to consider all 20^l peptide sequences to interpret an MS/MS spectrum, but could instead limit our search to peptides present in this database.

Currently, most proteins are identified by database search—effectively looking the answer up in "the back of a book." Indeed, an experimental spectrum can be compared with the theoretical spectrum for each peptide in such a database, and the entry in the database that best matches the observed spectrum usually provides the sequence of the experimental peptides. This forms the basis of the popular SEQUEST algorithm developed by John Yates and colleagues. Though the logic that SEQUEST used to determine whether a database entry matches an experimental spectrum is somewhat involved, the basic approach of the algorithms is just a linear search through the database.

One drawback to MS/MS database search algorithms like SEQUEST is that peptides in a cell are often slightly different from the "canonical" peptides present in databases. The synthesis of proteins on ribosomes is not the final step in a protein's life: many proteins are subject to further modifications that regulate protein activities and these modifications may be either permanent or reversible. For example, the enzymatic activity of some proteins is regulated by the addition or removal of a phosphate group at a specific residue. *Phosphorylation* is a reversible process: *protein kinases* add phosphate groups while *phosphatases* remove them. Proteins form complex systems necessary for cellular signaling and metabolic regulation, and are therefore often subject to a large number of biochemical modifications (e.g., phosphorylation and glycosylation).

In fact, almost all protein sequences are modified after they have been constructed from their mRNA template, and as many as 200 distinct types of modifications of amino acid residues are known. Since we are unable to predict these *post-translational modifications* from DNA sequences, finding naturally occurring modifications remains an important open problem.

Computationally, a chemical modification of the protein $p_1p_2...p_i...p_n$ at the position i results in increased mass of the N-terminal peptides $P_i, P_{i+1}, ..., P_n$ and increased mass of the C-terminal peptides $P_1^-, P_2^-, ..., P_{i-1}^-$. The computational analysis of modified peptides was also pioneered by John Yates, who suggested an exhaustive search approach that (implicitly) generates a virtual database of all modified peptides from a small set of potential modifications, and matches the experimental spectrum against this virtual database. It leads to a large combinatorial problem, even for a small set of modification types. The major difficulty with the Modified Protein Identification problem is that very similar peptides P_1 and P_2 may have very different spectra S_1 and S_2.

Our goal is to define a notion of spectral similarity that correlates well with sequence similarity. In other words, if P_1 and P_2 are a few modifications apart, the spectral similarity between S_1 and S_2 should be high. The shared peaks count is, of course, an intuitive measure of spectral similarity. However, this measure diminishes very quickly as the number of mutations increases, thus leading to limitations in detecting similarities by database search. Moreover, there are many correlations between the spectra of related peptides, and only a small proportion of these correlations is captured by the shared peaks count. The spectral convolution algorithm, below, reveals potential peptide modifications without an exhaustive search and therefore does not require generating a virtual database of modified peptides.

GREEDY ALGORITHMS

The algorithm *USChange* is an example of a greedy strategy: at each step, the cashier would only consider the largest denominator smaller than (or equal to) M. Since the goal was to minimize the number of coins returned to the customer, this seemed like a sensible strategy: you would never use five nickels in place of one quarter. A generalization of *USChange*, *BetterChange* also used what seemed like the best option and did not consider any others, which is what makes an algorithm "greedy." Unfortunately, *BetterChange* actually returned incorrect results in some cases because of its short-sighted notion of "good". This is a common characteristic of greedy algorithms: they often return suboptimal results, but take very little time to do so. However, there are lucky few greedy algorithms that find optimal rather than suboptimal solutions.

Genome Rearrangements

Waardenburg's syndrome is a genetic disorder resulting in hearing loss and pigmentary abnormalities, such as two differently coloured eyes. The disease was named after the Dutch ophthalmologist who first noticed that people with two differently coloured eyes frequently had hearing problems as ell.

In the early 1990s, biologists narrowed the search for the gene implicated in Waardenburg's syndrome to human chromosome 2, but its exact location remained unknown for some time. There was another clue that shed light on the gene associated with Waardenburg's syndrome, that drew attention to chromosome 2: for a long time, breeders scrutinized mice for mutants, and one of these, designated *splotch*, had pigmentary abnormalities like patches of white spots, similar to those in humans with Waardenburg's syndrome. Through breeding, the *splotch* gene was mapped to one of the mouse chromosomes. As gene mapping proceeded it became clear that there are groups of genes in mice that appear in the same order as they do in humans: these genes are likely to be present in the same order in a common ancestor of humans and mice—the ancient mammalian genome.

In some ways, the human genome is just the mouse genome cut into about 300 large genomic fragments, called *synteny blocks*, that have been pasted together in a different order. Both sequences are just two different shufflings of the ancient mammalian genome. For example, chromosome 2 in humans is built from fragments that are similar to mouse DNA residing on chromosomes 1, 2, 3, 5, 6, 7, 10, 11, 12, 14, and 17. It is no surprise, then, that finding a gene in mice often leads to clues about the location of the related gene in humans.

Every genome rearrangement results in a change of gene ordering, and a series of these rearrangements can alter the genomic architecture of a species. Analyzing the rearrangement history of mammalian genomes is a challenging problem, even though a recent analysis of human and mouse genomes implies that fewer than 250 genomic rearrangements have occurred since the divergence of humans and mice approximately 80 million years ago.

Every study of genome rearrangements involves solving the combinatorial puzzle of finding a series of rearrangements that transform one genome into another. The elementary rearrangement event in this scenario is the flipping of a genomic segment, called a *reversal*, or an *inversion*. One can consider other types of evolutionary events but in this book we only consider reversals, the most common evolutionary events. Biologists are interested in the most parsimonious evolutionary scenario, that is, the scenario involving the smallest number of reversals. While

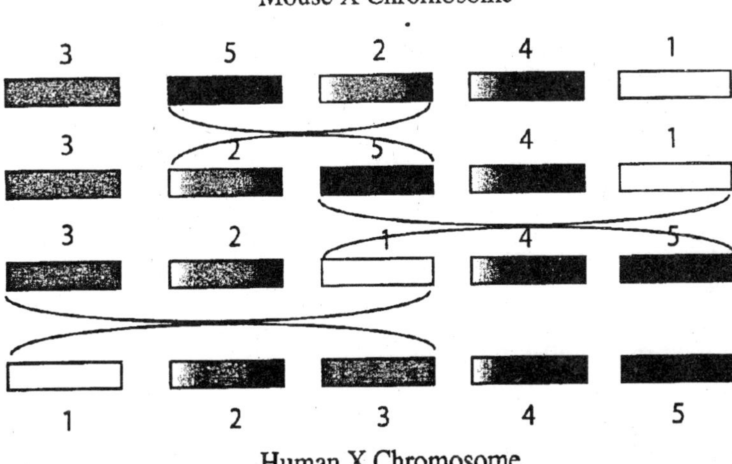

Fig. 8.18. Transformation of the mouse gene order into the human gene order on the X chromosome.

there is no guarantee that this scenario represents an actual evolutionary sequence, it gives us a lower bound on the number of rearrangements that have occurred and indicates the similarity between two species. Even for the small number of synteny blocks shown, it is not so easy to verify that the three evolutionary events represent a *shortest* series of reversals transforming the mouse gene order into the human gene order on the X chromosome. The exhaustive search technique that we presented in the else where would hardly work for rearrangement studies since the number of variants that need to be explored becomes enormous for more than ten synteny blocks. Below, we explore two greedy approaches that work to differing degrees of success.

Sorting by Reversals

In their simplest form, rearrangement events can be modeled by a series of reversals that transform one genome into another. The order of genes (rather, of synteny blocks) in a genome can be represented by a permutation $\pi = \pi_1\pi_2...\pi_n$. The order of synteny blocks on the X chromosome in humans is represented by (1, 2, 3, 4, 5), while the ordering in mice is (3. 5, 2, 4, 1). A reversal $\pi(i, j)$ has the effect of reversing the order of synteny blocks

$$\pi_i\pi_{i+1} \ldots \pi_{j-1}\pi_j$$

In effect, this transforms

$$\pi = \pi_1 \dots \pi_{i-1} \xrightarrow{\pi_i \pi_{i+1} \dots \pi_{j-1} \pi_j} \pi_{j+1} \dots \pi_n$$

Into

$$\pi \cdot \rho(i, j) = \pi_1 \dots \pi_{i-1} \xleftarrow{\pi_j \pi_{j-1} \dots \pi_{i+1} \pi_i} \pi_j \pi_{j+1} \dots \pi_n$$

For example, if π = 1243756, then π . ρ (3,6) = 1257346. With this representation of a genome, and a rigorous definition of an evolutionary event, we are in a position to formulate the computational problem that mimics the biological rearrangement process.

Reversal Distance Problem

Given two permutations, find a shortest series of reversals that transforms one permutation into another.

Input: Permutation π and σ.

Output: A series of reversals $\rho_1, \rho_2, \dots, \rho_t$ transforming π into σ (i.e., $\pi \cdot \rho_1 \cdot \rho_2 \dots \rho_t = \sigma$), such that *t* is minimum.

We call *t* the *reversal distance* between π and σ, and write *d*(π, σ) to denote the reversal distance for a given π and ρ. In practice, one usually selects the second genome's order as a gold standard, and arbitrarily sets σ to be the *identity permutation* 1 2 ... *n*. The sorting by Reversals problem is similar to the Reversal Distance problem, except that it requires only one permutation as input.

Sorting by Reversals Problem

Given a permutation, find a shortest series of reversals that transforms it into the identity permutation.

Input: Permutation π.

Output: A series of reversals $\rho, \rho_2 \dots, \rho_t$ transforming π into the identity permutation such that *t* is minimum.

In this case, we call *t* the *reversal distance* of π and denote it as *d*(π). When sorting, a permutation π = 1 2 3 6 4 5, it hardly makes sense to move the already-sorted first three elements of π. If we define *prefix*(π) to the number of already-sorted elements of π, then a sensible strategy for sorting by reversals is to increase *prefix*(π) at every step. This approach sort π in 2 steps: 1 2 3 <u>6 4</u> 5 → 1 2 3 4 <u>6 5</u> → 1 2 3 4 5 6. Generalizing this leads to an algorithm that sorts a permutation by repeatedly moving its ith element to the ith position.

```
Simple reversal sort (π)
for i ← 1 to n – 1
j ← position of element i in π (i.e, π_j = i)
if j ≠ i
π ← π. ρ(i, j)
output π
if π is the identity permutation
return
```

Simple reversal sort is an example of a greedy algorithm that chooses the "best" reversal at every step. However, the notion of "best" here is rather short-signted—simply increasing *prefix*(π) does not guarantee the smallest number of reversals. For example, *Simple reversal sort* takes five steps to sort 6 1 2 3 4 5:

612345 → 162345 → 126345 → 123645 → 123465 → 123456

However, the same permutation can be sorted in just two steps:

612345 → 543216 → 123456

Therefore, we can confidently say that *Simple reversal sort* is not a correct algorithm. In fact, despite its common sense appeal, *Simple reversal sort* is a terrible algorithm since it takes $n - 1$ steps to sort the permutation $\pi = n12 \ldots (n-1)$ even though $d(\pi) = 2$. Even before biologists faced genome rearrangement problems, compute scientists studied the related Sorting by Prefix Reversals problem, also know as the Pancake Flipping problem: given an arbitrary permutation π, find $d_{pref}(\pi)$, which is the minimum number of reversals of the form $\rho(1, i)$ sorting π.

The Pancake Flipping problem was inspired by the following "real-life" situation described by (the fictitious) Harry Dweighter: "The chief in our place is sloppy, and when he prepares a stack of pancakes they come out all different sizes. Therefore, when I deliver them to a customer, on the way to table I rearrangement them (so that the smallest winds up on top, and so on, down to the largest at the bottom) by grabbing several from the top and flipping them over, repeating this (varying the number I flip) as many times as necessary. If there are n pancakes, what is the maximum number of flips that I will eve have to use to rearrange them?"

An analog of *Simple reversal sort* will sort every permutation by at most $2(n - 1)$ prefix reversals. For example, one can sort 123645 by 4 prefix reversals (123645 → 632145 → 541236 →

321456 → 123456) but it is not clear whether there exists an even shorter series of prefix reversals to sort this permutation. William Gates, an undergraduate student at Harvard in the mid-1970, and Christos Papadimitriou, a professor at Harvard in the mid-1970s, now at Berkeley, made the first attempt to solve the problem and proved that any permutation can be sorted by at most $^5/_3(n + 1)$ prefix reversals. However, the Pancake Flipping problem remains unsolved.

Approximation Algorithms

We mentioned that, for many problems, efficient polynomial algorithms are still unknown and unlikely ever to be found. For such problems, computer scientists often finds a compromise in *approximation algorithms* that produce an approximate solution rather than an optimal one. The approximation ratio of algorithm A on input π is defined as $[A(\pi)]/[OPT(\pi)]$ where, $A(\pi)$ is the solution produced by the algorithm A and $OPT(\pi)$ is the correct optimal) solution of the problem. The *approximation ratio*, or *performance guarantee of algorithm A* is defined as its maximum approximation ratio over all inputs of size n, that is, as

$$\max_{|\pi|=n} = \frac{(\pi)}{OPT(\pi)}.$$

We assume that A is a *minimization* algorithm, i.e. an algorithm that attempts to minimize its objective function. For maximization algorithms, the approximation ratio is

$$\min_{|\pi|=n} = \frac{(\pi)}{OPT(\pi)}.$$

In essence, an approximation algorithm gives a worst-case scenario of just how far off an algorithm's output can be from some hypothetical perfect algorithm. The approximation ratio of *Simple reversal sort* is at least $(n - 1)/2$, so a biologist has no guarantee that this algorithm comes anywhere close to the correct solution. For example, if n is 1001, this algorithm could return a series of reversals that is as large as 500 times the optimal. Our goal is to design approximation algorithms with better performance guarantees, for example, an algorithm with an approximation ratio of 1 (by definition, a correct and optimal algorithm) would be the acme of perfection, but such algorithms can be hard to find.

Breakpoints: A Different Face of Greed

We have described a greedy algorithm that attempts to maximize *prefix*(π) in every step, but any chess player knows that greed often leads to wrong decisions. For example, the ability to take a queen in a single step is usually a good sign of a trap. Good chess players use a more sophisticated notion of greed that evaluates a position based on many subtle factors rather than simply on the face value of a piece they can take. The problem with *Simple reversal sort* is that *prefix*(π) is a naïve measure of our progress toward the identity permutation, and does not accurately reflect how difficult it is to sort a permutation. Below we define *breakpoints* that can be viewed as "bottlenecks" for sorting by reversals.

Using the number of breakpoints, rather than *prefix*(π), as the basis of greed leads to a better algorithm for sorting by reversals, in the sense that it produces a solution that is closer to the optimal one. It will be convenient for us to extend the permutation $\pi_1 ... \pi_n$ by $\pi_0 = 0$ and $\pi_{n+1} = n + 1$ on the ends. To be clear, we do not move π_0 or π_{n+1} during the process of sorting. We call a pair of neighbouring elements π_i and π_{i+1}, for $0 \leq i \leq n$, an *adjacency* if π_i and π_{i+1} are consecutive numbers; we call the pair of *breakpoint* if not.

The permutation has five adjacencies (21, 34, 45, 87, and 76) and four breakpoints (02, 13, 58 and 69). A permutation on n elements may have as many as $n + 1$ breakpoints (e.g. the permutation 061357248 on seven elements has eight breakpoints) and as few as 0 (the identity permutation 012345678). Every breakpoint corresponds to a pair of elements π_i and π_{i+1} that are neighbours in π but not in the identity permutation. In fact, the identity permutation is the only permutation with no breakpoints at all. Therefore, the nonconsecutive elements π_i and π_{i+1} forming a breakpoint must be separated in the process of transforming π to the identity, and we can view sorting by reversals as the process of eliminating breakpoints. The observation that every reversals can eliminate at most two breakpoints (one on the left end and another on the right end of

0 2 1 3 4 5 8 7 6 9

Fig. 8.19. Breakpoints, adjacencies, and strips for permutation 2 1 3 4 5 8 7 6 (extended by 0 and 9 on the ends).

the reversal) immediately implies that $d(\pi) \geq [b(\pi)/2$, where $b(\pi)$ is the number of breakpoints in π. The algorithm *Break point reversal sort* eliminates as many breakpoints as possible in every step in order to reach the identity permutation.

Breakpoint reversal sort(π)

```
while b(π) > 0
Among all reversals, choose reversal ρ minimizing b(π.ρ)
π ← π . ρ
output π
return
```

One problem with this algorithm is that it is not clear why *Breakpoint reversal sort* is a better approximation algorithm than *Simple reversal sort*. Moreover, it is not even obvious yet that *Breakpoint reversal sort* terminates! How can we be sure that removing some breakpoints does not introduce others, leading to an endless cycle? We define a *strip* in a permutation π as an interval between two consecutive breakpoints, that is, as any maximal segment without breakpoints. For example, the permutation 0213458769 consists of five strips: 0, 21, 345, 876 and 9. Strips can be further divided into *increasing* strips (345) and *decreasing* strips (21) and (876).

Single-element strips can be considered to be either increasing or decreasing, but it will be convenient to define them as decreasing (except for elements 0 and $n + 1$ which will always be classified as increasing strips). We present the following theorems, first to show that endless cycles of breakpoint removal cannot happen, and then to show that the approximation ratio of the algorithm. While the notion of "theorem" and "proof" might seem overly format for what is, at heart, a biological problem it is important to consider that we have modeled the biological process in mathematical terms. We are providing analytically that the algorithm meets certain expectations. This notion of proof without experimentation is very different from what a biologist would view as proof, but it is just as important when working in bioinformatics.

Theorem. If a permutation π contains a decreasing strip, then there is a reversal ρ that decreases the number of breakpoints in π, that is, $b(\pi \,.\, \pi) < b(\pi)$.

Proof. Among all decreasing strips in π, choose the strip containing the smallest element k (k = 3 for permutation

012 765 8 43 9). Element $k - 1$ in π cannot belong to a decreasing strip, since otherwise we would choose a strip ending at $k - 1$ rather than a strip ending at k. Therefore, $k - 1$ belongs to an increasing strip; moreover, it is easy to see that $k - 1$ terminates this strip (for permutation 012 765 8 43 9, $k - 1 = 2$ and 2 is at the right end of the increasing strip 012). Therefore elements k and $k - 1$ correspond to two breakpoints, one at the end of the decreasing strip ending with k and the other at the end of the increasing strip ending in $k - 1$.

Reversing the segment between k and $k - 1$ brings them together, as in 0127658439 → 01234 8 567 9, thus reducing the number of breakpoints in π. For example *Breakpoint reversal sort* may perform the following four steps when run on the input (0827651439) in order to reduce the number of breakpoints :

(0 8 2 765 1 4 3 9)	$b(\pi) = 6$
(0 2 8765 1 4 3 9)	$b(\pi) = 5$
(0 234 1 56789)	$b(\pi) = 3$
(0 4321 56789)	$b(\pi) = 2$
(0123456789)	$b(\pi) = 0$

In this case, *Breakpoint reversal sort* steadily reduces the number of breakpoints in every step of the algorithm. In other cases, (e.g., the permutation (01 567 234 89) without decreasing strips), no reversal reduces the number of breakpoints. In order to overcome this, we can simply find any increasing strip (excluding π_0 and π_{n+1}, of course) and flip it. This creates a decreasing strip and we can proceed.

Improved breakpoint reversal sort (π)

```
while b(π) > 0
if π has a decreasing strip
Among all reversals, choose reversal ρ minimizing b(π.ρ)
else
choose a reversal ρ that flips an increasing strip in π.
π → π.ρ
output π
return
```

The theorem below demonstrates that such "no progress" situation do not happen too often in the course of *Improved breakpoint reversal sort*. In fact, the theorem quantities exactly how often those situations could possibly occur and provides an approximation ratio guarantee.

Theorem. Improved breakpoint reversal sort is an approxi-mation algorithm with a performance guarantee of at most 4.

Proof. Theorem implies that as long as π has a decreasing strip, *Improved breakdown reversal sort* reduces the number of breakpoints in π. On the other hand, it is easy to see that if all strips in π are increasing, then there might not be a reversal that reduces the number of breakpoints. In this case *Improved breakpoint reversal sort* finds a reversal ρ that reverses an increasing strip(s) in π. By reversing an increasing strip, ρ creates a decreasing strip in π implying that *Improved breakpoint reversal sort* will be able to reduce the number of strips at the next step. Therefore, for every "no progress" step, *Improved breakpoint reversal sort* will make progress at next step which means that *Improved breakpoint reversal sort,* eliminates at least one breakpoint in every two steps. In the worst-case scenario, the number of steps in *Improved breakpoint reversal sort* is at most $2b(\pi)$ and its approximation ratio is at most

$$\frac{2b(\pi)}{d(x)}.$$

Since

$$d(\pi) \geq \frac{b(\pi)}{2},$$

Improved breakpoint reversal sort has as performance guarantee bounded above by

$$\frac{2b(\pi)}{d(\pi)} \leq \frac{2b(\pi)}{\frac{b(\pi)}{2}} = 4$$

Greedy Approach to Motif Finding

We saw a brute force algorithm to solve the Motif Finding problem. With a disappointing running time of $O(l \cdot n^t)$, the practical limitation of that algorithm is that we simply cannot run it on biological samples. We choose instead to rely on a faster greedy technique, even though it is not correct and does not result in an algorithm with a good performance guarantee.

Despite the fact that this algorithm is an approximation algorithm with an unknown approximation ratio, a popular tool based on this approach developed by Gary Stormo and Gerald

Hertz in 1989, *Consensus*, often produces results that are as good as or better than more complicated algorithms. *Greedy motif search* scans each DNA sequence only once. Once we have scanned a particular sequence *i*, we decide which of its *l*-mer has the best contribution to the partial alignment score *Score* (*s*, *i*, *DNA*) for the first *i* sequences and immediately claim that this *l*-mer is part of the alignment. The pseudocode is shown below.

```
Greedy motif search (DNA, t, n, l)
bestMotif ← (1, 1, ..., 1)
s ← (1, 1, ..., 1)
for s_1 ← 1 to n − l + 1
for s_2 ← 1 to n − l + 1
if Score (s, 2, DNA) > Score (bestMotif, 2, DNA)
BestMoti f_1 ← s_1
BestMoti f_2 ← s_2
s_1 ← Best Moti f_1
s_2 ← BestMoti f_2
for i ← 3 to t
for s_i ← 1 to n − l + 1
if Score (s, i, DNA ) > Score (BestMotif, i, DNA)
bestMoti f_i ← s_i
s_i ← bestMoti f_i
return bestMotif
```

Greedy motif search first finds the two closest *l*-mers—in the sense of Hamming distance—in sequences 1 and 2 and forms a 2 × l *seed matrix*. This stage requires $l(n - l + 1)^2$ operations. At each of the remaining $t - 2$ iterations *Greedy motif search* extends the seed matrix into a matrix with one more row by scanning the *i*th sequence (for $3 \leq i \leq t$) each of the remaining $t - 2$ sequences and selecting the one *l*-mer that has the maximum *Score* (*s*, *i*). This amounts to roughly $l \cdot (n - l + 1)$ operations in each iteration. Thus, the running time of this algorithms is $O(ln^2 + lnt)$, which is vastly better than the $O(ln^t)$ of *Simple motif search* or even the $O(4^l nt)$ of *Brute force medians string*.

When *t* is small compared to *n*, *Greedy motif search* really behaves as $O(ln^2)$, and the bulk of the time is actually spent locating the *l*-mers from the first two sequences that are the most similar. As you can imagine, because the sequences are scanned sequentially, it is possible to construct input instances where *Greedy motif search* will miss the optimal motif. One important difference between the popular *Consensus* motif finding software tool and the algorithm presented here is that *Consensus* can scan the sequences in a random order, thereby making it more difficult to construct inputs that elicit worst-case behaviour. Another important difference is that *Consensus* saves a large number (usually at least 1000) of seed matrices at each iteration rather than only the one that *Greedy motif search* saves, making *Consensus* less likely to miss the optimal solution. However, no embellishment of this greedy approach will be guaranteed to find an optimal motif.

9

Models of Voltage Gating

In this chapter we will learn how to model current flowing through ion channels and how to include the feedback regulation of the channel behavior by voltage. But let us be very clear about what we are modeling in this chapter. Only in the most idealized sense are we modeling the behavior of excitable cells like neurons. In fact, we will be creating *point models* of patches of membrane, with only implicit consideration through the use of ion concentrations of the idea that the membrane encloses a compartment. We do not consider at all that a neuron is a highly complicated cell that integrates signals within *dendrites* and a *soma* before producing *action potentials* in an excitable *axon*.

Electrophysiology is the study of ionic currents and electrical activity in cells and tissues. Because this field has its roots in classical physics, traditionally it has been the most quantitative field in cell physiology. The groundbreaking work of the physiologists Hodgkin and Huxley and others in elucidating the mechanism of action potentials in the squid giant axon before and after the Second World War was the first major breakthrough of dynamical modeling in physiology. In the latter half of the twentieth century, the introduction of the patch-clamp technique established firmly that ionic currents are carried by proteins that act as gated ionic pores. More recently, genetic engineering techniques have been employed to clone, modify, and characterize the gating mechanisms of many types of channels.

In this chapter we focus on voltage gated ionic currents. We begin by reviewing the basic concepts of electrical behavior in cells. Next, we describe classical activation and inactivation kinetics and how the voltage clamp technique can be used to study these currents. We use the Morris-Lecar model for action potentials in the giant barnacle muscle to illustrate how voltage gated channels can interact to produce oscillations and action potentials. The Morris-Lecar model is nonlinear but involves only two variables. With only two variables, we can analyze the dynamics of the equations for this model using phase plane techniques. For completeness, we close with brief introductions to the Hodgkin-Huxley model of the squid giant axon and FitzHugh-Nagumo models.

Table 9.1. Important Definitions in Electrophysiology

Definition	*Abbreviation*	*Value*
Avogadro's number	N	$6.02 \cdot 10^{23}$/mol
Faraday's constant	F	$9.648 \cdot 10^{4}$C/mol
elementary charge	e	$1.602 \cdot 10^{-19}$ C
gas constant	R	8.315 J/(mol . K)
joule	J	1 V . C
volt	V	1 J/C
ampere	A	1 C/s

Basis of Ionic Battery

The electrical behavior of cells is based upon the transfer and storage of charge. We are used to thinking about electricity as the movement of electrons, but current can be carried by any charged particle, including ions such as K+, Na^+, and Ca^{2+} in solution. As we will see, the ability of cells to generate electrical signals is entirely dependent on the evolution of ion-specific pumps and pores that allow the transfer of charge up and down gradients. Ion pumps use energy in the form of ATP to transport ions against a concentration gradient.

Before we begin our discussion of the ionic battery, it is usefull to recall several important definitions from elementary physics listed in Table 9.1. An ion's *valence* is the number of charges, plus or minus, that it carries. A given number of *divalent* Ca^{2+} ions would carry twice the amount of charge as the same number of *univalent* K^+ ions.

Using these relationships, we see that the transfer of 1 mole of K^+ ions in a period of one second would carry a current equal to Faraday's constant. It requires one joule of energy (*i.e.* ATP consumed by ion pumps) to separate one coulomb of charge across one volt of potential (which is the definition of volt).

Nernst Potential: Charge Balances Concentration

Biological fluids such as cytoplasm and extracellular fluid contain numerous ions. Consider the case where the two ions K^+ and any monovalent anion A^- are in solution such that the concentration is different across the impermeable membrane but the two ions are equal in concentration on the same side of the membrane. Before we make any changes, there is no potential difference across the membrane because the charge between the K^+ ions and the A^- ions is balanced on each side due to the equivalent concentrations. If we insert a nons-elective pore into the membrane, concentration and charge equilibrate such that there are equal concentrations of each ion on both sides of the membrane, and the voltage across the membrane is again zero.

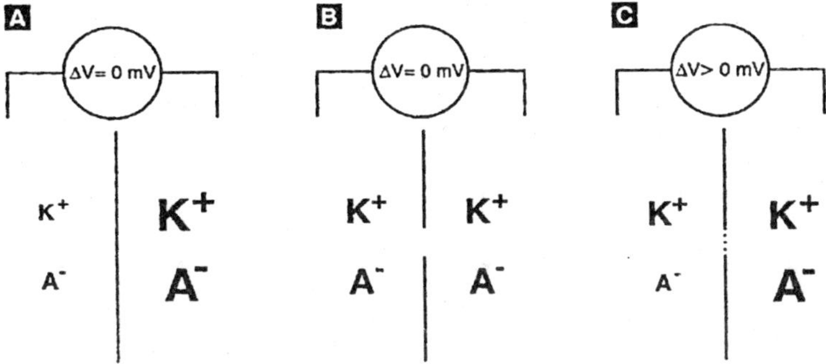

Fig. 9.1. (A) Concentration and charge are balanced on each side of the membrane, so there is no ΔV across the membrane. (B) Due to the nonselective pore, charge and concentration are balanced everywhere, and so there is no ΔV across the membrane. (C) A K^+ selective pore allows K^+ but not A^- to pass through the membrane.

It is when we insert into the membrane an ion-selective pore that allows only the passage of K^+. Because $[K^+]$ is greater on one side of the membrane, K^+ ions diffuse through the K^+ pore down the concentration gradient. Because the membrane is not permeable to the anion A^-, each K^+ ion that passes down the concentration gradient carries a positive charge that is not balanced by an accompanying A^-. Because the transfer of these charges establishes an electrical potential gradient, K^+ ions continue to move from high concentration to low concentration until the growing force due to the electrical potential difference is balanced by the (opposite) force generated by the concentration difference.

The *equilibrium potential,* where the electrical and osmotic forces are balanced, is given by the *Nernst equation*. The Nernst equation is derived from the expression for the change in Gibbs free energy when one mole of an ion of valence z is moved across a membrane:

$$\Delta G = -RT \ln \frac{[\text{ion}]_{\text{out}}}{[\text{ion}]_{\text{out}}} + \Delta VFz. \qquad ...(1)$$

At equilibrium, ΔG is zero. Rearranging gives us the Nernst potential:

$$\Delta G = V_{\text{nernst}} = \frac{RT}{zF} \ln \frac{[\text{ion}]_{\text{out}}}{[\text{ion}]_{\text{in}}}$$

$$= 2.303 \frac{RT}{zF} \log_{10} \frac{[\text{ion}]_{\text{out}}}{[\text{ion}]_{\text{in}}}$$

where R and F are given in Table 9.1, T is temperature (in kelvin), and z is the valence of the ion as previously defined. At body temperature, RT/F is approximately 60 mV. Therefore, a 10-fold difference in the concentration of a monovalent ion like K^+ would result in approximately –60 mV of potential difference across a membrane ($[K^+]$ is greater inside a neuron). Because the Nernst potential represents the equilibrium of the thermodynamic system, the potential difference evolves to that given by the Nernst equation regardless of the initial starting potential. This tendency for the system to move toward the equilibrium potential is the basis of the *ionic battery* used in the modeling of electrophysiological phenomena. In electrophysiology, the equilibrium potential is called the *reversal potential,* because departure from that point of zero current flux results in the positive or negative flow of ions.

Resting Membrane Potential

The Nernst potential is the equilibrium potential for one permeant ion. In reality, no channel is perfectly selective for a given ion, and there are various channels selective for various ions in a given cell as well! The Goldman-Hodgkin-Katz (GHK) equation is related to the Nernst equation, but considers the case where there are multiple conductances. The GHK equation determines the resting membrane potential of a cell from a weighted sum of the various conductances:

Table 9.2. Resting ion concentrations

Ion	*Cytoplasmic Concentration (mM)*	*Extracellular Concentration (mM)*
Squid Giant Axon		
K+	400	20
Na+	50	440
Cl-	40	560
Mammalian Neuron		
K+	135	3
Na+	18	145
Cl^-	7	120

$$V_{GHK} = \frac{RT}{F}\ln\frac{P_K[K^+]_{out} + P_{Na}[Na^+]_{out} + P_{Cl}[Cl^-]_{in}}{P_K[K^+]_{out} + P_{Na}[Na^+]_{in} + P_{Cl}[Cl^-]_{out}} \quad ...(3)$$

where P_i is the relative permeability for ion *i*, which must be determined experimentally. While it looks like a straightforward extension of the Nernst equation, the GHK equation requires assumptions about both the interaction of ions and their ability to diffuse within channels.

We are not going to deal with these complicating details here, and in addition, we will assume perfect selectivity for our ion channels. If we define the *conductance*, $g = 1/R$ as the reciprocal of the resistance, we can use a similar weighted-sum formalism for calculating the membrane potential of a cell:

$$V_m = \frac{\Sigma_i(V_i \cdot g_i)}{\Sigma_i g_i}, \quad ...(4)$$

where V_i is the Nernst-equation-derived reversal potential for ion *i* calculated using (2). For example, the membrane potential for a cell containing Na^+, K^+, and Cl^- ions would be

$$V_m = \frac{(V_{Na \cdot g_{Na}}) + (V_K \cdot g_K) + (V_{Cl} \cdot g_{Cl})}{g_{Na} + g_K + g_{Cl}} \quad ...(5)$$

This is the expression for the resting membrane potential that we will use throughout the remainder of the book. Because the resting membrane potential is the weighted average of Nernst potentials for the various ions, the ion with the greatest permeability contributes the most.

The Nernst potential represents an equilibrium between electrical and osmotic forces on an ion across a membrane: It is what the membrane potential would be if a particular ion were at equilibrium across a membrane. Each ion has its own characteristic Nernst potential given by equation (2), which is determined by its concentration gradient across the membrane. The actual membrane potential is some "average" of the Nernst potentials of the several ions that are able to cross the membrane; each ion contributing to the actual membrane potential according to the permeability (or conductance) of the membrane for that particular ion. The resting potential across a membrane given by (5) is a steady state in which the currents carried by different ions across the membrane cancel each other out. The GHK potential is a steady state solution of a different kinetic equation for total current, based on different assumptions about how ions cross the membrane.

Membrane Model

We know from Ohm's law that current flows down a voltage gradient in proportion to the resistance in the circuit. Current is therefore expressed as

$$I = \frac{V}{R} = gV. \qquad ...(6)$$

To a first approximation, our task in modeling electrophysiological phenomena is to describe how the conductance of the membrane to various ions changes with time and then to keep track of the changes in current and voltage that result.

The conceptual idea behind contemporary electrophysiological models originates in the work of K.S. Cole, who pioneered the notion that cell membranes could be likened to an electronic circuit. Cole's basic circuit elements are (1) the phospholipid bilayer, which acts as a capacitor in that it accumulates ionic charge as the electrical potential across the membrane changes; (2) the ionic permeabilities of the membrane, which act as resistors in an electronic circuit; and (3) the electrochemical driving forces, which act as batteries driving the ionic currents. This analogy to electrical circuits is now widely relied upon for developing models of electrical activity in membranes.

Equations for Membrane Electrical Behavior

Given the several conductances and their reversal potentials, we calculate using (5) what the membrane potential will be after

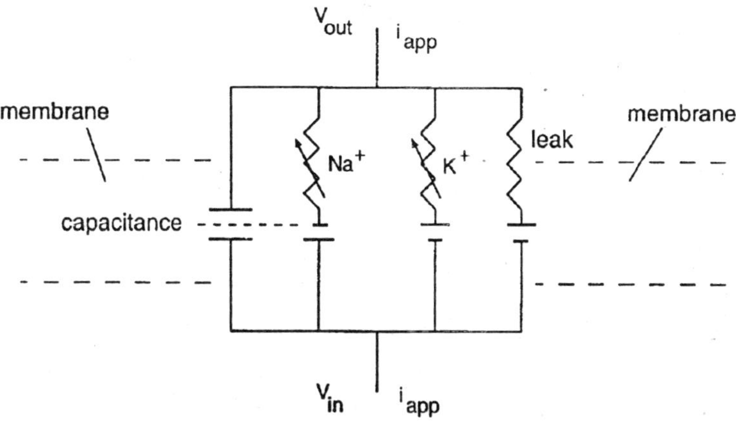

Fig. 9.2. The equivalent electrical circuit for an electrically active membrane.

the system has stabilized. This equations for the resting membrane potential tell us nothing about how the system evolves to the steady state. Because we are interested in the time course of the membrane voltage, we have to study the dynamics of the various currents that flow in and out of the cell. We can approximate the current flow through a single K^+ channel using Ohm's law and an assumption that the reversal potential stays constant:

$$I_K = -g_K(V - V_K). \qquad ...(7)$$

Here g_K is the conductance of the K^+ channel and the leading negative sign is necessary because of our definitions for "in" and "out." V_K is the K^+ reversal potential determined by the Nernst equation, and $V - V_K$ represents the driving force across the membrane provided by the ionic battery. We assume that the reversal potential for a given ion remains constant, which is equivalent to assuming that restorative mechanisms such as ionic pumps can keep pace with electrical activity on a time scale that prevents the ionic battery from running down. This is a reasonable assumption for a large cell, which would have a small surface area to volume ratio. In a small cell, with a large surface to volume ratio, the ion transfer necessary to change the membrane potential might have a large effect on the intracellular ionic concentration and thus the strength of the ionic battery.

Of course, numerous ions are responsible for the electrical behavior in a cell, and the total current is the sum of the individual ionic currents.

$$\Sigma I_i = \Sigma{-g_i}(V - V_i) = -g_K(V - V_K) - g_{Na}(V - V_{Na}) - \ldots \quad \ldots(8)$$

To translate the electric circuit diagram into ODEs, we use the traditional interpretation of each circuit element along with Kirchoffs law. Assuming that the membrane acts as a capacitor, the capacitive current across the membrane can be written

$$I_{cap} = C\frac{dV}{dt}, \quad \ldots(9)$$

where C is the capacitance of the membrane and V is the membrane potential, defined as the electrical potential difference between the inside and outside of the cell. Kirchoff's law dictates that capacitive current must balance with the ionic current and any currents that might be applied, say, through experimental manipulation. This implies that

$$I_{cap} = I_{ion} + I_{app}, \quad \ldots(10)$$

where the sum is over all the ionic currents. Using the expressions in (8) – (10) this can be rewritten

$$C\frac{dV}{dt} = \sum_i g_i(V - V_i) + I_{app}. \quad \ldots(11)$$

To solve this differential equation for voltage, we must know how the gated conductances g_i depend on V (and possibly time). In general, the g_i will not be linear functions of V, and therefore the problem is to find the time and possible voltage dependence of the various conductances.

Activation and Inactivation Gates

Channels can be thought to *have gates* that regulate the permeability of the pore to ions. These gates can be controlled by membrane potential, producing *voltage gated* channels; by chemical ligands, producing *ligand gated* channels; or by a combination of factors. In a series of experiments, Alan Hodkgin, Andrew Huxley, and others established experimentally the voltage dependence of ion conductances in the electrically excitable membrane of the squid giant axon. Hodgkin and Huxley shared a Nobel Prize with John Eccles in 1963 for work on nerve signaling.

Models of Voltage-Dependent Gating

The mathematical description of voltage-dependent activation and inactivation gates is based on the mechanism.

$$C \underset{k^-}{\overset{k^+}{\rightleftharpoons}} O, \quad \ldots(12)$$

What distinguishes a voltage-dependent gating mechanism from a passive mechanism is the voltage dependence of the rate constants. Recall the fraction of open channels *fo* satisfies the differential equation

$$\frac{dfo}{dt} = \frac{-(f_O - f_\infty)}{\tau}, \qquad \text{...(13)}$$

where

$$f_\infty = \frac{k^+}{k^+ + k^-} \text{ and } \tau = \frac{1}{k^+ + k^-}. \qquad \text{...(14)}$$

Because ionic channels are composed of proteins with charged amino acid side chains, the potential difference across the membrane can influence the rate at which the transitions from the open to closed state occur. According to the Arrhenius expression for the rate constants, the membrane potential V contributes to the energy barrier for these transitions:

$$k^+ \propto \exp\left(\frac{-\Delta V^+}{RT}\right) \text{ and } k^+ \propto \exp\left(\frac{-\Delta V^-}{RT}\right). \qquad \text{...(15)}$$

The rate constants will have the form

$$k^+ = k_o^+ \exp(-\alpha V) \text{ and } k^- = k_o^- \exp(-\beta V), \qquad \text{...(16)}$$

where k_o^+ and k_o^- are independent of V. Substituting the relationships in (16) into the expressions for f_∞ and τ and rearranging, we obtain

$$f_\infty = \frac{1}{1 + k_o^-/k_o^+ \exp((\alpha - \beta)V)}, \qquad \text{...(17)}$$

$$\tau = \frac{1}{k_o^+ \exp(-\alpha V)} \cdot \frac{1}{1 + k_o^-/k_o^+ \exp((\alpha - \beta)V)}. \qquad \text{...(18)}$$

We can define

$$S_o = \frac{1}{\beta - \alpha} \qquad \text{...(19)}$$

and

$$V_o = \frac{\ln(k_o^-/k_o^+)}{\beta - \alpha} \qquad \text{...(20)}$$

If we substitute S_o and V_o into (17) and (18), we have

$$f_\infty = \frac{1}{1 + \exp(-(V - V_o)/S_o)} \qquad \text{...(21)}$$

$$\tau = \frac{\exp(\alpha V)}{k_o^+} \cdot \frac{1}{1+\exp(-(V-V_o)/S_o)}. \quad ...(22)$$

Finally, both of these expressions can be rewritten in terms of hyperbolic functions:

$$f_\infty = 0.5(1 + \tanh((V - V_o)/2S_o)), \quad ...(23)$$

$$\tau = \frac{\exp(V(\alpha+\beta)/2)}{2\sqrt{k_o^+ k_o^-}\cosh((V-V_o)/2S_o)}. \quad ...(24)$$

Recall that f_∞ gives the fraction of channels open at equilibrium at the membrane potential V. Thus for a fixed value of V it gives the open fraction after transient changes in f_o have damped out with a characteristic time τ.

An activation gate tends to open and an inactivation gate tends to close when the membrane is depolarized. Whether a gate activates or inactivates with depolarization is determined by the sign of S_o: a positive sign implies activation and a negative sign inactivation. This is illustrated in Figure 9.3A, where the dependence of f_∞ on V has been plotted for an activation gate with $V_0 = -25$ mV and $S_0 = 5$ mV and an inactivation gate with $V_o = -50$ mV and $S_o = -2$ mV.

Notice that the magnitude of S_0 determines the steepness of the dependence of f_∞ on V, whereas the value of V_0 determines the voltage at which half of the channels are open. The dependence of τ on V for these activation and inactivation gates is illustrated in Figure 9.3B (assuming that $\alpha = -\beta$ and $2\sqrt{k_o^+ k_o^-} = 0.2\ \mathrm{ms}^{-1}$). When $\alpha = -\beta$ and $\phi = 1/\left(2\sqrt{k_o^+ k_o^-}\right)$,

$$\tau = \frac{\phi}{\cosh((V-V_o)/2S_o)}. \quad ...(25)$$

Voltage Clamp

In order to measure the voltage across a cell membrane or the current flowing through a membrane, microelectrodes are inserted into cells. These electrodes can be used both to measure current and voltage and to apply external current. In order to measure the voltage dependence of the activation and inactivation of ion conductances, a technique called the *voltage clamp* is used. This is an electronic feedback device that adjusts the applied current I_{app} to match and counter the membrane currents such that the membrane voltage is held constant. To see what this accomplishes, consider a membrane with a single gated ionic

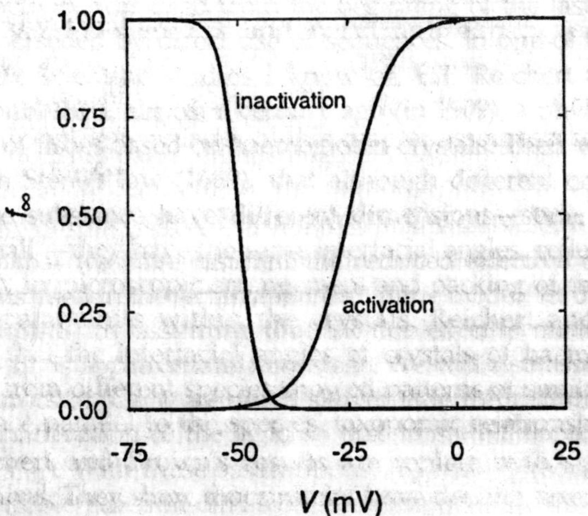

B

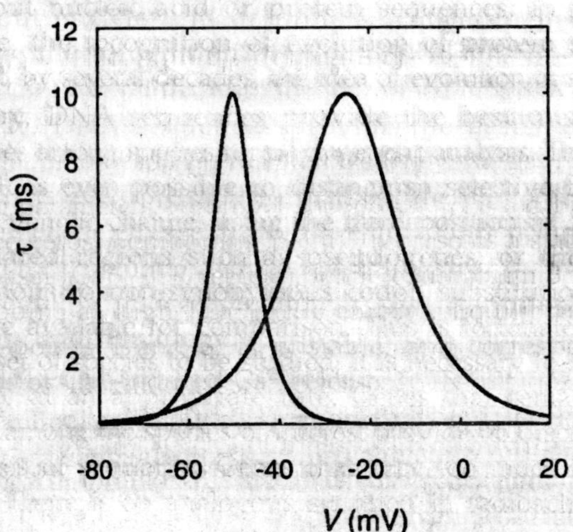

Fig. 9.3. (A) Equilibrium open fractions (f_∞) for an inactivation gate (V_0 = –50 mV and S_0 = –2 mV) and activation gate (V_0 = – 25 mV and S_0 = –5 mV) as a function of voltage. (B) The characteristics relaxation times τ for the activation and inactivation gates in (A) as a function of voltage, which are peaked around the values of V_0 and have a width determined by S_0.

current. If we assume that the total conductance is the result of the activation of many channels, the conductance g that we have used above can be denned as the product of the maximum possible conductance $\bar{g}$ and the fraction of open channels f_o that we have already encountered.

$$g - f_o\bar{g}. \quad \text{...(26)}$$

Table 9.3. Consistent Electrical Units

Name (Symbol)	*Units*	*Abbreviation*
voltage *(V)*	10^{-3} volt	mV
time *(t)*	10^{-3} second	ms
conductance *(g)*	10^{-9} Siemens	nS
capacitance *(C)*	10^{-12} farad	pF
current *(I)*	10^{-12} ampere	pA

We can include this new relationship in the differential equation for membrane potential that we have already seen:

$$C\frac{dV}{dt} = -f_o\bar{g}(V - V_{\text{rev}}) + I_{\text{app}}, \quad \text{...(27)}$$

where V_{rev} is the reversal potential given by the Nernst equation. If we can apply a current that is equal and opposite to the current flowing through the membrane,

$$I_{\text{app}} = \bar{g}\xi(V - V_{\text{rev}}), \quad \text{...(28)}$$

then the right-hand side of (27) is zero and the voltage must be constant. Because V is constant, $\xi = f_o$ and the time dependence of the applied current comes only from the dependence of f_o on t as determined by the gating equation:

$$\frac{d\xi}{dt} = \frac{-(f_O - f_\infty)}{\tau}. \quad \text{...(29)}$$

Thus the time dependence of the applied current provides a direct measurement of the gated current at a fixed voltage. Note that throughout the remainder of the text we will drop the overbar with the understanding that conductances g_i refer to maximum conductances to be scaled by gating variables.

To carry out a voltage clamp measurement like this it is necessary to block all but a single type of current. While this is not always possible, specific toxins and pharmacological agents have proven useful. For example, tetrodotoxin (TTX) from the puffer fish selectively blocks voltage gated Na^+ currents.

It is not difficult to simulate a voltage clamp measurement using (28) – (29). However, to carry out either an experimental measurement or a simulation, a consistent set of electrical units must be used. As we have seen, the standard unit for membrane potential is millivolts (mV), and because the characteristic times for voltage-dependent gates *t* are in milliseconds (ms), this is taken as the standard unit of time. Currents are typically expressed in $\mu A/cm^2$ and capacitances as $\mu F/cm^2$. For a typical cell of area 10^{-6} cm^2, this translates to a whole-cell current of picoamperes (1 pA = 10^{-12} A) and a whole cell capacitance of picofarads (1 pF = 10^{-12} F). Cellular dimensions are usually reported in micrometers ("microns"), and there are 10^{-8} square centimeters per square micron. Because most biological channels have a conductance *g* on the order of 1 to 150 pS, whole-cell conductances of nanosiemens are usually expressed in mS/cm^2, because the units of $V - V_{rev}$ are mV. An alternative consistent set of units uses current in femtoamperes (1 fA = 10^{-15} A), conductance in pS, and capacitance in fF.

To simulate a voltage clamp experiment we have solved (29) and plotted the resulting current given by (28). The simulation shown in Figure 9.4A represents a typical set of experiments in which the membrane potential is clamped at a holding potential (−60 mV), then changed to various test potentials for a fixed interval (40 ms), and finally returned to the holding potential. The value of the holding potential generally is chosen so that there is little or no current through the channel. This greatly simplifies the interpretation of the current at the test voltages. The increase in current when the potential is clamped at the test values is governed by the exponential increase in f_O with characteristic time $\tau(V_{test})$. When the potential is clamped again at the holding potential, the resulting current is called the *tail current*. Its decline is also exponential, but because $V = -60$ mV during this'period, the characteristic time is now $\tau(-60 \text{ mV})$.

According to (28) it can be expressed as $I = gf_\infty(V - V_{rev})$. Thus for an activating current like that in the simulations, when V is large enough, $f_\infty \approx 1$ and the current is a linear function of V. The curvature at lower voltages is caused by the shape of the activation function f_∞. In the jargon of circuit theory, currents like this are said to *rectify*. The delay in the onset of the maximum current, which is determined by the value of τ, has led to channels like the one being referred to as *delayed rectifiers*.

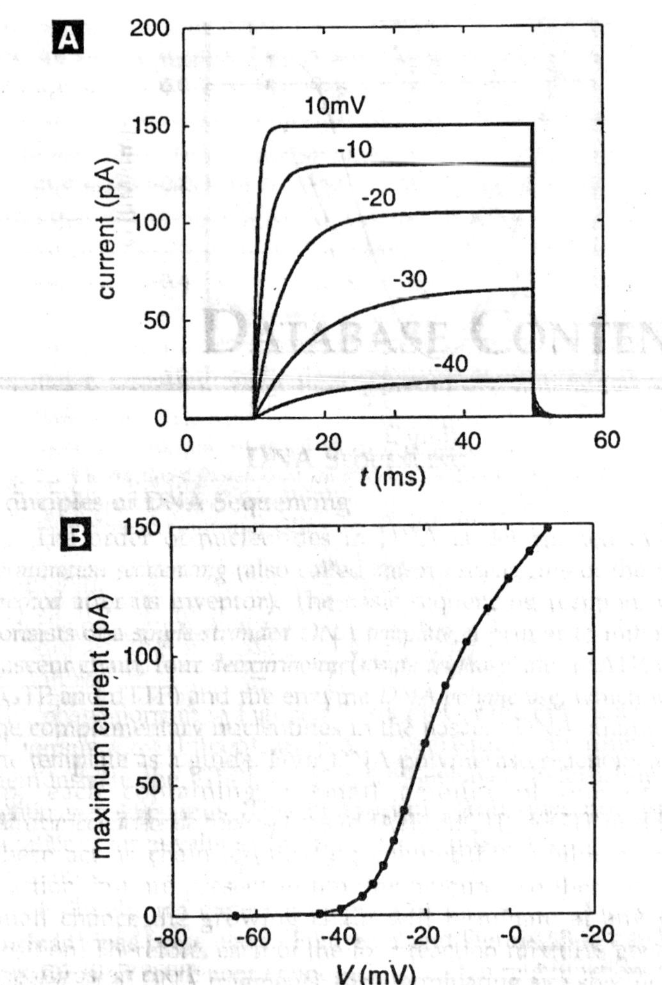

Fig. 9.4. Simulation of voltage clamp experiment. (A) Current records resulting from 40 ms depolarizations from the holding potential of –60 mV to the indicated test potentials. (B) The maximum current as a function of test potential taken from records like those in (A).

Interacting Ion Channels: Morris-Lecar Model

Application of a depolarizing current to barnacle muscle fibers produces a broad range of electrical activity. Careful experimental work by a number of research groups has indicated that the giant barnacle muscle fiber contains primarily voltage gated K^+ and Ca^{2+} currents along with a K^+ current that is activated by intracellular Ca^{2+}, a so-called K^+_{Ca} current. Neither of the voltage gated currents shows significant inactivation in voltage clamp experiments. The trains of depolarization-induced action potentials must occur via a mechanism different from that proposed by Hodgkin and Huxley for the squid giant axon, which, as we will see later, include channel inactivation.

Morris and Lecar proposed a simple model to explain the observed electrical behavior of the barnacle muscle fiber. Their model involves only a fast activating Ca^{2+} current, a delayed rectifier K^+ current, and a passive leak. They tested the model against a number of experimental conditions in which the interior of the fiber was perfused with the Ca^{2+} chelator EGTA in order to reduce activation of the Kca current. Their simulations provide a good explanation of their experimental measurements. The model translates into two equations:

$$C\frac{dV}{dt} = -g_{Ca}m_{\infty}(V - V_{Ca}) - g_K w(V - V_K) - g_L(V - V_L) + I_{app}, \quad ...(30)$$

$$\frac{dw}{dt} = \frac{\phi(w_{\infty} - w)}{\tau}. \quad ...(31)$$

Here m_{∞} is the fraction of voltage-dependent Ca^{2+}channels open, and this is a function of voltage but not time. Furthermore, w is the fraction of open channels for the delayed rectifier K^+ channels, and the conductances g_L. g_{Ca}, and g_K are for the leak, Ca^{2+}, and K^+ currents, respectively. We use w rather than the previously used *fa* for the fraction of open channels for historical reasons. The functions

$$m_{\infty} = 0.5[1 + \tanh((V - v_1)/v_2)], \quad ...(32)$$

$$w_{\infty} = 0.5[1 + \tanh((V - v_3)/v_4)], \quad ...(33)$$

$$\tau = 1/\cosh((V - v_3)/(2 \cdot v_4)), \quad ...(34)$$

are the equilibrium open fractions for the Ca^{2+} current and the K^+ current, and the activation time constant for the delayed rectifier. Again note that *m* is not a dynamic variable. The reason for this is that we have assumed that the time constant for *m* is short enough that *m* is always in steady state, $m = m_{\infty}$. The idea of fast and slow processes is arguably one of the most important concepts in modeling.

We have solved the Morris-Lecar equations for four values of the applied current I_{app} and plotted the time series for *V*. Current in the Morris-Lecar model is specified in $\mu A/cm^2$, however to simplify notation somewhat we will assume that the

Table 9.4. Morris-Lecar oscillator parameters (Type II)

Parameter	*Value*
C	20μF/cm^2
V_K	–84 mV
g_K	8 mS/cm^2
V_{Ca}	120 mV
g_{Ca}	4.4 mS/cm^2
V_{leak}	–60 mV
g_{leak}	2 mS/cm^2
$v1$	–1.2 mV
$v2$	18 mV
$v3$	2mV
vA	30 mV
ϕ	0.04/ms

cell has a total surface area of 10^{-6} cm^2, so that μA/cm^2 corresponds to 1 pA of total current. In the absence of applied current the equations have a stable steady state near –60 mV. Although increasing I_{app} to 60 pA produces a brief transient action potential, the effect of the depolarization simply produces a steady state near –35 mV. Depolarization with a current of 150 pA, on the other hand, produces a steady train of action potentials reminiscent of those observed experimentally. In the presence of depolarizing currents much greater than this, the simulated barnacle cell can no longer sustain continuous spiking, as shown at I_{app} = 300pA.

Phase Plane Analysis

The mechanistic features underlying continuous spiking and action potentials can be understood easily using phase plane analysis. Because of the ease of representation in two dimensions, two variable models such as the Morris-Lecar model are particularly amenable to this technique. Phase plane analysis is a powerful way to determine how the behavior of a system will change with changes in the various parameters in a system. Several types of plots are utilized as part of what is generically called phase plane analysis:

1. A *phase portrait* consists of the variables describing a system plotted against each other rather than as a function of time to produce a *trajectory* in *phase space*. A phase portrait tells us how the variables interact for a given set of parameters.

2. A *vector field* shows us the direction in which a system will evolve from any location in phase space.
3. *Nullclines* are plotted in phase space, and show us the values of a pair of variables at which one of the variables does not change. In other words, for a coupled system of equations *X(x, y, t)* and *Y(x, y, t)* nullclines are the solutions to the equations

$$\frac{dX}{dt} = 0, \frac{dY}{dt} = 0 \qquad ...(35)$$

Note that there is a nullcline for each variable. The points of intersection of two nullclines are particularly interesting points in phase space that we will discuss below.

Phase plane analysis plots for the Morris-Lecar model are shown in Figure 9.5B and Figure 9.5C. Figure 9.5B shows trajectories, nullclines, and the vector field for I_{app} = 150 pA, which

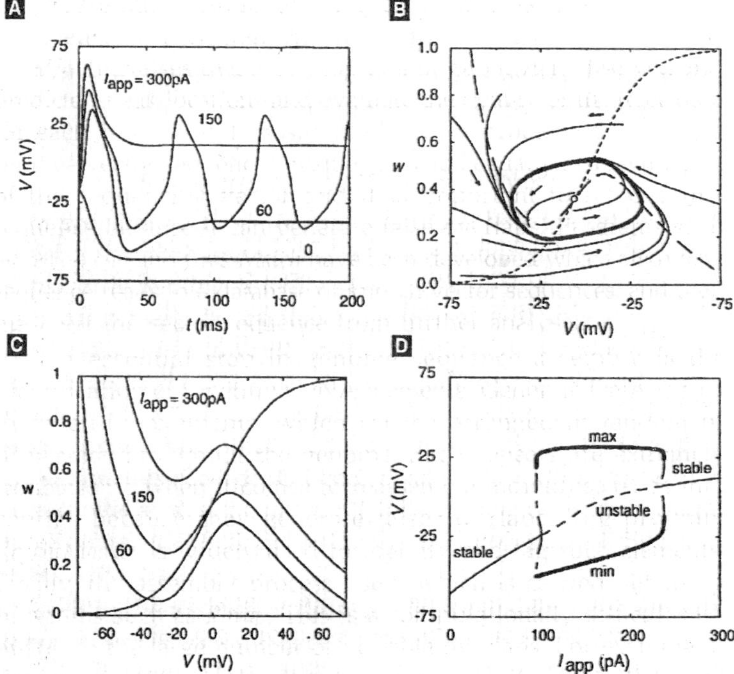

Fig. 9.5. (A) Voltage simulation for the Morris-Lecar equations using the indicated applied currents and parameters as in Table 9.4. (B) The phase plane for the Morris-Lecar model for I_{app} = 150 pA. (C) Nullclines for several values of I_{app}. (D) A bifurcation diagram that catalogues the dynamical states of the Morris-Lecar model as a function of I_{app} with the other parameters fixed.

leads to the pattern of repetitive spiking. The phase portrait consists of a number of representative trajectories together with a unique, closed trajectory indicated by the heavy line. This trajectory is called a *stable limit cycle* because it is the cyclic curve to which all the neighboring trajectories converge no matter where in phase space they originate. The trajectories circulate around the steady state in a counterclockwise direction as indicated by the velocity vector field, which is shown as small arrows.

The nullclines are indicated by the lighter segmented lines. The *V*-nullcline, which has the inverted "N" shape, is given by the long-dashed line. It is the solution to

$$0 = -g_{Ca}m_{\infty}(V - V_{Ca}) - g_K w(V - V_K) - g_L(V - V_L) + I_{app} \quad ...(36)$$

for each value of *V*.

The *w*-nullcline is the solution to

$$0 = \frac{\phi(w_{\infty} - w)}{\tau} \quad ...(37)$$

for each value of *V*.

Note that the limit cycle circulates around the intersection of the two nullclines. We saw that different values of I_{app} lead to different behaviors and that I_{app} = 150 pA was the only value of the four tested that results in oscillations. Changing the value of I_{app} affects the nullclines of the system. We see that increasing I_{app} from 60 pA to 150 pA and then to 300 pA raises the *V* nullcline but leaves the *w* nullcline unaltered. If we examine carefully, we notice something interesting: As we saw above, the limit cycle obtained when I_{app} = 150 pA circulates around the intersection of the nullclines at that parameter value. At I_{app} = 300 pA the system evolves to a steady state that corresponds to the intersection of the nullclines obtained for that I_{app}. The same is true for I_{app} = 60 pA. We know that a point on a nullcline corresponds to a point at which the variable of interest is not changing, and so it makes sense that the intersection of nullclines represents a combination of variables for which the system as a whole does not change. The intersection of the nullclines for I_{app} = 60 pA and I_{app} = 300 pA are examples of *fixed points,* and represent *stable steady states.* What is different about the intersection of nullclines obtained when I_{app} =150 pA?

Stability Analysis

Fixed points can be either stable or unstable and the intersection of nullclines obtained when I_{app} = 150 pA is an unstable fixed point. Notice that the two trajectories that start nearest to the intersection of the nullclines diverge away from the intersection and toward the limit cycle. The trajectory of the system is thus driven away from the intersection of the nullclines while at the same time being constrained to orbit around it. The existence of a stable limit cycle should be no more surprising than the existence of stable steady states. It corresponds to a closed trajectory to which all neighboring trajectories converge. The tools provided for phase plane analysis and stability analysis in some ODE packages make it easy to determine steady states, limit cycles, and stability, and therefore how a system like the Morris-Lecar model will behave.

Eigenvalues describe the solution of the linear system: Negative real eigenvalues indicate a stable solution, positive real, eigenvalues indicate an unstable solution, and complex eigenvalues indicate the presence of oscillations in the solution. Stability analysis for nonlinear systems is not quite so straightforward, in that we can only determine the stability in such a system in a very small region around the a fixed point. In essence we linearize the system, either analytically (with a *Taylor series*) or numerically (with a software package). We then use the same tools that we used for the linear system in the small linearized region around the fixed point in the nonlinear system.

This book emphasizes the use of computational tools rather than mathematical analysis. An intuitive understanding of the concept of stability can be obtained without mathematical rigor. However, it must be stressed that there are myriad subtleties that can be appreciated only with an understanding of the underlying mathematics. Other excellent sources covering this material on a reasonably introductory level are available, and a particularly detailed analysis of the firing proterties of the Morris-Lecar system has been published.

The simulations show that a stable limit cycle occurs for the Morris-Lecar model only for certain values of I_{app}, such as I_{app} = 150 pA. We can determine the stability of the fixed points corresponding to different values of I_{app} using a numerical package. The intersection of the nullclines for I_{app} = 150 pA indicated by the open circle in Figure 9.5B is an unstable steady state at $V = -.460$ and $w = 0.459$, a fixed point with eigenvalues $\lambda\pm = 0.264$ and 0.033 (values are rounded). It is unstable because the eigenvalues are positive real numbers, and yet these parameters result in stable oscillations. This is an excellent example of how the local stability of a fixed point does not tell the whole story for a nonlinear system. Unlike the case with linear systems, here positive eigenvalues can be associated with stable oscillations in phase space around the unstable fixed point.

Table 9.5. Fixed points and their eigenvalues

I_{app}	V	W	*Eigenvalue*	*Eigenvalue*
0 pA	–60.855 V	0.015	–0.037	–0.096
60 pA	–37.755V	0.070	–0.055 + /0.063	–0.055 – /0.063
110 pA	–19.219V	O.T96	0.055 + (0.045	0.055 – /0.045
150 pA	–0.460 V	0.459	0.264	0.033
180 pA	6.656 V	0.577	0.025 + /0. 139	0.025 –/0.139
300 pA	14.302V	0.694	–0.137 + /0.1 17	–0.137 –/0.1 17

Using a systematic stepping procedure, we can test the stability of the system over a wide range of parameter values. Table 9.5 shows the eigenvalues for the values of I_{app} that we have examined plus some additional intermediate values. As I_{app} is increased from I_{app} = 0, the eigenvalues are seen to change from negative (stable node) to complex with negative real parts (stable focus), to complex with positive real parts (unstable focus) as oscillations emerge. The oscillations at I_{app} = 150 correspond to eigenvalues that are both real and positive (unstable node). As I_{app} is increased further, the eigenvalues become complex again with positive real parts, and eventually become complex with negative real parts as oscillations cease. Fixed points at which the qualitative character of the solution changes with a change in a parameter are called *bifurcation* points.

By making even smaller changes in I_{app} and testing the stability of the steady state, it is possible to locate the specific values at which the stability of the steady state changes. This procedure could become very laborious, and so numerical algorithms have been developed that allow investigation of bifurcations to be done automatically. In the bifurcation diagram, the characteristic values of the membrane potential are plotted on the ordinate as a function of I_{app}. The thin full and dashed lines are the steady-state values of V for each value of I_{app}, with the full lines representing stable steady states and the dashed lines unstable states. The points I_{app} & 94 pA and 212 pA are bifurcation points where the stability of the steady state changes. Near these points two new dynamical features appear: a stable limit cycle and an unstable limit cycle. This type of bifurcation is called a *subcritical Hopf bifurcation*. Note that while the detail is not apparent, the unstable limit cycle turns back and then coalesces at a turning point bifurcation with the stable limit cycle at a value of I_{app} that is smaller than the bifurcation point, and therefore there is a small region of bistability. The bifurcation diagram also records the maximum and minimum values of V on the limit cycles with heavy full lines. For example, at I_{app} = 150 pA the points on the heavy line correspond to the maximum and minimum of the spikes.

Why Do Oscillations Occur?

If the following three conditions on the Morris-Lecaf equations hold, then oscillations will occur:

1. the V nullcline has the inverted "N" shape;
2. a single intersection of the V– and w-nullclines occurs between the maximum and minimum of the "N;"
3. the rate of change of V is much greater than w.

All three conditions are met for the parameter values in Table 9.4, giving rise to oscillations. The importance of the slow change in w, *i.e.*, the "delay" of the delayed rectifier, can be seen by examining the trajectories. If the rate of change of w were large with respect to V, then the trajectories would not depolarize and hyperpolarize rapidly as they do. This is tested by decreasing the value of the parameter ϕ, thereby increasing the value of the characteristic time for relaxation of w.

By altering the time scale we see why oscillations occur when the rate of change of V is very much faster than w. In this case, we can treat changes in V under the assumption that w is constant, and we need only to consider the voltage equation with w fixed:

$$C\frac{dV}{dt} = -g_{Ca}m_{\infty}(V - V_{Ca}) - g_K w(V - V_K) - g_L\,(V - V_L) + I_{app} \quad \text{....(38)}$$

Because only the voltage is changing on this time scale, we can examine its dynamical behavior using the one-dimensional phase portrait, rather than a phase plane. This is shown in Figure 9.6A, where the total current, which is proportional to the rate of change of V, is plotted along with the Ca^{2+} and K^{+} currents for w

= 0.35 and I_{app} = 150 pA. The total current vanishes at three points, which are the steady states for the voltage when w is fixed. States 1 and 3 are stable, and state 2 is unstable, as indicated by the velocity vectors in the figure. Note that at state 1 the membrane is polarized and the outward K^+ current dominates the Ca^{2+} inward current, whereas the opposite holds true at state 3. As we shall see, the oscillations can be thought of as transitions back and forth between the polarized and depolarized states, driven by slow changes in activation of the delayed rectifier current.

Figure 9.6A also explains why the V-nullcline in the Morris-Lecar model has the inverted "N" shape: The K^+ and leak currents exceed the inward Ca^{2+} current at polarized voltages between states 1 and 2, whereas the Ca^{2+} current exceeds the other currents between states 2 and 3. This voltage-dependent competition between inward and outward currents leads to a maximum and minimum in total current and therefore the inverted "N" shape for the nullcline.

The three steady states of the voltage also can be found graphically in the phase plane of Figure 9.6B by locating the intersection of the line w = 0.35 with the V-nullcline. It is clear from the figure that if w exceeds 0.468 (the maximum on the right branch of the V-nullcline), then the voltage has only a single polarized steady state (intersection with w) on the far left branch of the V-nullcline. Similarly, if w is smaller than 0.251 (the

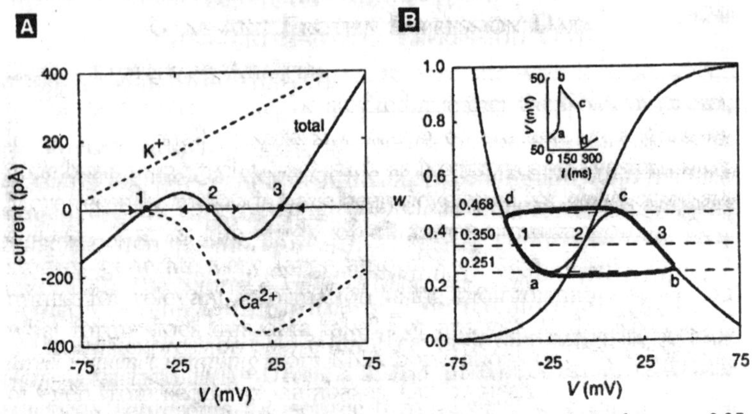

Fig. 9.6. (A) The K^+, Ca^{2+}, and total current ($I_{Ca} + I_k + I_{leak} - I_{app}$) when w = 0.35. Stage 1 and 3 are stable steady states, and state 2 is unstable as indicated by the velocity vectors. (B) The phase plane for the Morris-Lecar model for I_{app} = 150 pA, except that $\tau(V)$ has been increased by a factor of ten.

minimum on the left branch of the V-nullcline), then the voltage has only a single depolarized steady state on the right branch. For $0.251 \le w \le 0.468$ two stable steady states and one unstable state occur, and the voltage is said to be *bistable*.

To understand how bistability on the fast time scale (neglecting w) leads to oscillations in the complete system, we need to understand how w changes on the longer time scale. Assume that initially, the membrane is polarized at state 1, with $w = 0.35$. Because $w = 0.35$ is above the w-nullcline at point 1, w will decrease. As w decreases, V will stay close to the V-nullcline because it relaxes rapidly to the closest steady-state value. The trajectory thus follows the polarized branch, as indicated by the heavy line, until the minimum at $w = 0.251$ is reached. Beyond the minimum (near *a*), stable polarized states no longer exist, and V rapidly relaxes to the only remaining steady state (near *b*), which is on the depolarized, far right branch of the V-nullcline. During the depolarization, however, the w-nullcline is crossed. Thus on the depolarized branch, w increases and tracks the V-nullcline upward until the maximum at $w = 0.468$ is reached (near *c*) and the membrane rapidly repolarizes to the polarized branch (near *d*).

The abrupt transitions from the polarized to depolarized branch and back again have led to the name *relaxation oscillator* for systems of equations that have well-separated time scales. For the parameters, the Morris-Lecar model is not a relaxation oscillator. However, when the characteristic time for w is increased by a factor of 10 (by decreasing ϕ), the limit cycle closely approximates that for a relaxation oscillator. If the characteristic time were increased sufficiently, or the characteristic time for V were decreased sufficiently by altering the capacitance, then the trajectory would coincide with the bistable portions of the V-nullclines, and the rapid excursions of the voltage would occur precisely at $w = 0.251$ and 0.468. The inset shows a single voltage spike that illustrates the rapid upstroke and downstroke for the limit cycle. The inset also illustrates that the shape of the depolarized and polarized portions of the spike reflects the shape of the two branches of the V-nullcline.

Excitability and Action Potentials

Another dynamical feature of the Morris-Lecar model is *excitability*. A useful working definition of excitability is that a

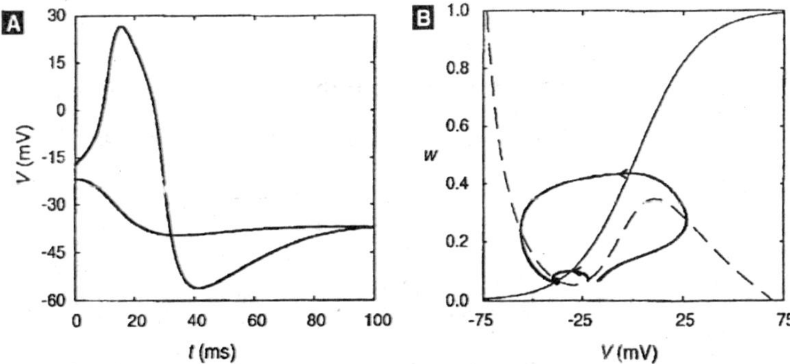

Fig. 9.7. Excitability in the Morris-Lecar model for I_{app} = 60 pA. (A) An initial deviation of the voltage to –22 mV relaxes rapidly to the steady-state voltage, whereas a deviation to –17 mV produces an action potential. (B) The trajectories in (A) represented in the phase plane.

system is excitable when small perturbations return to the steady state, but larger (*i.e.*, above a threshold) perturbations cause large transient deviations away from the steady state. An example of this is shown in Figure 9.7A, where the time course of the voltage with an applied current of 60 pA and initial conditions of $w(0) = w^{ss} = 0.070$ and either $V(0) = -22$ or –17 mV are used. For subthreshold $V(0)$ such as –22 mV, the voltage increases only slightly before it decreases monotonically to its steady state-value of about –37 mV. For suprathreshold deviations, like that for $V(0) =$ –17 mV, the voltage increases dramatically, producing an action potential spike before returning to steady-state.

The explanation for excitability can be understood most easily in the phase plane. In Figure 9.7B we have plotted the trajectories for the two initial conditions in Figure 9.7A along with the two nullclines. The trajectories for subthreshold initial conditions like $V(0) \leq -22$ mV start at points in phase space *above* the *V*-nullcline and *below* the w nullcline. This implies that the initial velocity vector, and therefore the initial trajectory, points in the direction of smaller potentials and larger values of the activation of the delayed rectifier K^+ current. Thus the voltage begins to decrease and continues to do so because the K^+ current activates as w increases. This contrasts with the trajectory that starts at $V(0) = -17$ mV, for which the initial velocity vector points in the direction of increasing voltage. Even though w is .increasing in this region, which is also below the w-nullcline, the rate of increase of voltage exceeds that of w, and the trajectory moves to higher voltages until it crosses the V-nullcline and begins to decrease.

The threshold value of $V(0)$ above which action potentials occur depends on the shape of the nullclines and the rate of activation of w (altered by ϕ). However, it is close to the point where a line drawn parallel to the V axis at w^{ss} crosses the V-nullcline. It is not hard to check that it will approach this point as w is made to change more slowly than V.

Type I and Type II Spiking

Many investigators are now interested in applying biophysical models of neurons and other spike-generating mechanisms to the study of information transfer using information-theoretic measures. It is particularly important, therefore, that the dynamical behavior of a model be characterized over the space of parameters and that the characteristics of the spike generating model match the intended use. Both the Hodgkin-Huxley model, described below, and the Morris-Lecar model produce trains of action potentials with commonly used parameter sets and when sufficiently depolarized. Muscle fiber and axons innervating muscle fiber are examples of situations where a strong, consistent signal is required. In our investigations of the Morris-Lecar model, we have seen that depolarization beyond threshold results in oscillations that begin at a characteristic nonzero frequency. This behavior has been defined as a Type II oscillator.

Many neurons exhibit firing properties that are fundamentally different from those of the Type II oscillator. Models of these cells can produce arbitrarily low frequencies of oscillations. These models are classified as Type I oscillators. The Morris-Lecar model can exhibit Type I or Type II behavior, depending on the parameters that are chosen, and the Hodgkin-Huxley model also can be modified to show this behavior. See Table 9.6 for the Morris-Lecar Type I parameters that differ from Type II parameters. The Morris-Lecar Type II oscillator might not be appropriate for the study of subtle aspects of information transfer or intra-neuronal coupling. We have seen that Type II spiking results from a subcritical Hopf bifurcation as input current is increased. Type I spiking results from a saddle-node bifurcation. We also note that there are other levels of complexity for spiking and oscillating models. Models of bursting cells, which

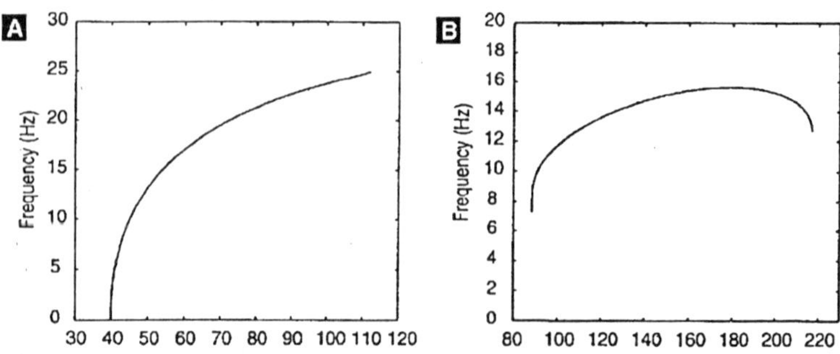

Fig. 9.8. Example of voltage-frequency plots for the Morris-Lecar model with parameters that result in (A) type I dynamics and (B) type II dynamics.

exhibit trains of oscillations separated by periods of quiescence, require additional slow variables.

Table 9.6. Morris-Lecar type I oscillator parameters

Parameter	*Value*
g_{Ca}	4 mS/cm^2
$v3$	12 mV
$v4$	17.4 mV
ϕ	0.066/ms

Hodgkin-Huxley Model

Ordinarily, a discussion of membrane electrophysiology might begin with the famous model of the squid giant axon developed by Hodgkin and Huxley (the HH model). Our goal here has been to facilitate a clear mathematical understanding of oscillatory behavior in systems of ion channels. While it remains a seminal accomplishment in the history of physiology, the HH model is complicated and not amenable to the phase plane methods of analysis that we have used to understand dynamical electrical behavior in cells. The student should be familiar with the HH model, for historical reasons and because it is still widely used. We cannot cover the work of Hodgkin and Huxley in the detail it deserves, but the concepts learned in the study of the Morris-Lecar oscillator will aid the student in understanding the HH model. We recommend that the student consult Bertil Hille's *Ionic Channels of Excitable Membranes*, which contains one of the best treatments of the HH model and the history behind it. "Hille" is also the best consolidated source of information about the biophysics of ion channels, and it should be familiar to any person attempting to model electrical behavior.

The HH model is empirical. Many voltage clamp experiments were performed by Hodgkin and Huxley, and their data were fit to expressions that they incorporated into the model without consideration of an underlying mechanism for the channel gates. One of the remarkable aspects of Hodgkin and Huxley's work is that their model was developed without a molecular understanding of the mechanism. In fact, it required almost thirty years of intensive research after their work to formulate a realistic cartoon of the mechanisms underlying the ionic currents in cells. Although Hodgkin and Huxley justified these expressions on empirical grounds, it is possible to derive the gating expression used in the Hodgkin-Huxley model using mechanistic models.

From their voltage clamp and other measurements, Hodgkin and Huxley deduced that the sodium conductance involved two voltage-dependent "gates," an *activation* gate and an *inactivation*, gate and that the potassium conductance had a single activation gate. Note that the presence of an inactivation gate differs from the Morris-Lecar model. To account for these facts they represented the ionic conductances in the following form:

$$g_{NA} = \bar{g}_{Na} m^3 h, \qquad ...(39)$$

$$g_K = \bar{g}_K n^4 \qquad ...(40)$$

where the terms $\bar{g}$ represent maximal conductances, and m and n are the activation gating variables and h the inactivation. The exponents on m and n were chosen for the best fit to experimental data. These gating variables were postulated to satisfy linear differential equations where the variables "relax" to voltage-dependent values, e.g., m_∞, that vary between zero and one with voltage-dependent time constants, *e.g.*, τ_m.

Putting the ODEs for the gating variables together with (11) gives the primary equations for the Hodgkin-Huxley model:

$$CdV/dt = \bar{g}_{Na} m^3 h(V - V_{Na}) - \bar{g}_K n^4 (V - V_K), - \bar{g}_{leak}(V - V_{leak}) + I_{app}, \qquad ...(41)$$

$$dm/dt = -(m - m_\infty)/\tau_m, \qquad ...(42)$$

$$dh/dt = -(h - h_\infty)/\tau_h, \qquad ...(43)$$

$$dn/dt = -(n - n_\infty)/\tau_n. \qquad ...(44)$$

Hodgkin and Huxley added the third conductance $\bar{g}_{leak}$ to their voltage equation to account for a small voltage-independent conductance that they attributed to a "leak" in the membrane,

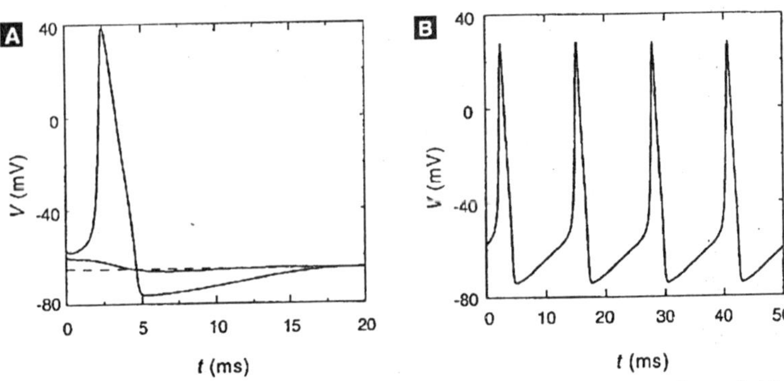

Fig. 9.9. (A) The solution of the Hodgkin-Huxley equations for three different initial values of the membrane potential and no applied current. (B) Continuous spiking occurs under the same conditions with an applied current I_{app} = 15.

possibly through their microelectrode. The nonlinear terms in (41) are obvious in the activation and inactivation gates. Not so obvious are the nonlinear-ides in (42)-(44). However, all the voltage-dependent terms in those equations are nonlinear functions of V as well.

The student should solve the HH equations to provide a basis for comparison and further exploration. The equations for the Hodgkin-Huxley spike generator are readily available from other sources, and it is not difficult to run simulations with a package for solving ODEs. One must be careful to choose a suitable numerical method of solution due to the nonlinear equations. The *Runge-Kutta* method has been used with a time step of 0.05, which is appropriate because reducing the time step to 0.01 gives no noticeable change in the results.

Figure 9.9A shows calculations with I_{app} = 0, but with $V(0)$ = –65, –60, and –57 mV. The steady state for the voltage with these parameters is clearly –65 mV; however, if the initial value of V exceeds about –59 mV, then the equations produce an action potential spike, and we have another example of a Type II oscillator. The Hodgkin-Huxley equations can also produce repetitive firing of action potentials. The parameters in that simulation are identical to those in Figure 9.9A, except that I_{app} = 15.

FitzHugh-Nagumo Class Models

We have seen that a fast variable with a cubic nullcline for voltage and a slower variable with a monotonically increasing nullcline for channel opening are sufficient to produce oscillations. In some cases, and particularly for mathematical analysis, it may be beneficial to completely abstract the mathematical equations for an oscillator from the underlying physical processes in order to understand general properties of these systems. Because two-variable systems are useful for the phase plane methods we have discussed, we can contrive such a system.

The two differential equations will be functions of both state variables:

$$\frac{dv}{dt} = f(v, w),$$

$$\frac{dv}{dt} = g(v, w). \qquad ...(45)$$

By definition, the nullcline is obtained by setting the right-hand side of the equation to zero and plotting the function in the (v, w) phase plane. We can create the behavior we desire by specifying the shapes of $f(v, w)$ and $g(v, w)$. The slow variable, w, is easiest, because we wantg (v, w) to be similar to a line describing the increase of v with w. Note that v here is the dependent variable for the equation for a line:

$$g(v, w) : v = \gamma w, \qquad ...(46)$$

and therefore

$$\frac{dw}{dt} = v - \gamma w, \qquad ...(47)$$

where γ is the slope parameter.

The fast variable v will have a form similar to (3.47) with w as the dependent variable:

$$f(v, w): w = f(v), \qquad ...(48)$$

and therefore

$$\frac{dv}{dt} = f(v) - w. \qquad ...(49)$$

Because we know that w needs to be a cubic function of v, let us try

$$f(v) = Bv(v - \beta(\delta - v), \qquad ...(50)$$

where A is a parameter to scale the amplitude of the curve. The equations for the complete system are therefore

$$\frac{dv}{dt} = Av(v - \beta)(\delta - v) - Cw,$$

$$\frac{dw}{dt} = \varepsilon(v - \gamma w). \qquad ...(51)$$

The parameter ε has been added to more easily control the speed of one variable relative to the other, and the parameter C affects the coupling strength. For an appropriate choice of all of the parameters, we would expect that this system of abstracted differential equations would produce oscillations.

This dynamical paradigm appeared first in cellular neuroscience in simplifications of the Hodgkin-Huxley equations by FitzHugh (1961), and later as part of an independent development by Nagumo et al. (1962). In fact, the various parameters we have contrived have physical interpretations in the context of a simplification of the Hodgkin-Huxley circuit diagram we discussed earlier. This type of system, involving a linear nullcline for the slow variable and a cubic nullcline that has the inverted "N" shape for the fast variable, are given the generic name *FitzHugh-Nagumo* (or FH-N) models.

INDEX